An Illustrated Guide to Pollen Analysis

P. D. Moore, B.Sc., Ph.D.
and
J. A. Webb, B.Sc.
Department of Plant Sciences,
King's College, London

HODDER AND STOUGHTON

LONDON · SYDNEY · AUCKLAND · TORONTO

British Library Cataloguing in Publication Data
Moore, Peter Dale
    An illustrated guide to pollen analysis.
    – (Biological science texts).
    1. Palynology – Europe 2. Pollen, Fossil
    3. Paleobotany – Quaternary
    I. Title II. Webb, J A III. Series
    561'.13   QE993.2

    ISBN 0-340-17236-3
    ISBN 0-340-21449-X   Pbk

Microscope Slides

Sets of microscope slides of named pollen types are
available from Philip Harris Biological Limited,
Oldmixon, Weston-super-Mare, Avon BS24 9BJ. The
types included in the sets have been carefully selected
by the author to complement this book. Further
details available on request.

ISBN 0 340 17236 3 Boards
ISBN 0 340 21449 X Unibook

First printed 1978

Printed and bound in Great Britain for
Hodder and Stoughton Educational,
a division of Hodder and Stoughton Limited,
Mill Road, Dunton Green, Sevenoaks, Kent, by
Richard Clay (The Chaucer Press) Ltd,
Bungay, Suffolk.

BIOLOGICAL SCIENCE TEXTS

General Editor
Don R. Arthur, M.Sc., Ph.D., D.Sc., F.I.Biol.
Professor of Zoology, King's College,
University of London

Other books in the series:

PHOTOSYNTHESIS
G. E. Fogg, Sc.D., F.R.S.

CELL RESPIRATION
W. O. James, F.R.S.

PLANT MINERAL NUTRITION
E. J. Hewitt, B.Sc., Ph.D., D.Sc., F.I.Biol., A.K.C.
T. A. Smith, B.Sc., Ph.D., M.I.Biol.

PATTERNS OF CHANGE IN TROPICAL PLANTS
G. P. Chapman, B.Sc., Ph.D.

LIGHT AND LIFE
L. O. Björn

Foreword

This book provides an excellent introduction to the study of pollen grains and spores from higher plants and from cryptogams respectively. It is suitably designed for the use of both University undergraduates and for sixth-formers in school. Its simplicity of style, its magnificently lucid approach and remarkable freedom from scientific jargon make it eminently readable. The authors are at pains to point out what can be achieved by a study of pollen grains and spores while recognizing the limitations. They discuss in the first chapter the relevant applications of pollen analysis in respect of the interpretation of the natural phenomena of the past and how these data may be useful in interpreting the evolution of the contemporary environment. In terms of the present environments the importance of palynology in the public health sector is appreciated and on which more fundamental research seems to be necessary, particularly in relation to pollen grain and spore dispersion. The situation in respect of the environment in which pollen grains are preserved is examined in the second chapter and serves to illustrate the ever-increasing challenge to cross-fertilize ideas between scientific disciplines to achieve greater understanding of extinct and extant natural phenomena. The collection of samples, structure and identification of pollen grains and spores, is expectedly a large element in a book of this sort. It is more readily understandable by the inclusion of an adequate glossary, a key and a wealth of illustrations. Such information, a necessary requisite for practical work and for carrying out project work, is clearly expressed and laid out. Methods of pollen counting and of pollen diagram construction leading to analytical procedures are described, while the very important feature on interpretation of such data is discussed at some length. The limitations affecting the validity or otherwise of interpretations based on the use of stereoscan electron-microscopy, of numerical taxonomy and of computer techniques are also examined and seem a profitable field for further investigation. The potential of pollen analysis as a tool in the further understanding of our environment is clearly indicated in this book, as for example in studying the pollen content of the atmosphere in respect of allergies. Nevertheless, the authors are also well aware of the fact that more research needs to be done in this area of study, if only to identify pollen grains below the genus or family level. From this point of view the value of this book lies in the insight it may provide for further advanced studies. Dr. P. D. Moore and Miss J. A. Webb are to be congratulated on producing this volume, for it serves an important and developing need and will be of very considerable value in both schools and universities.

D.R.A.

Preface

This book seeks to do a number of things. We hope that it will serve as a laboratory manual which will provide a brief description of the main techniques involved in the recovery of fossil pollen grains from different types of recent sediment, such as peats and lake deposits. In the key we hope to have supplied workers with a means for the identification of the vast majority of the pollen and spore types which one is likely to encounter in north-west European, Quaternary deposits. In addition, we have added to the key a series of photographs which should assist in this identification process. We have considered the errors inherent in palynological techniques and data and have reviewed the methods currently available to reduce these errors to a minimum. Finally we have discussed how data should be interpreted and the considerations which should be borne in mind constantly while this task is being undertaken.

Although Chapter 1 is concerned with the application of palynology in a wide variety of fields, we should stress that our main concern has been with the analysis of pollen in recent (Quaternary) geological strata and its interpretation in terms of surrounding vegetation and hence environment. The key will undoubtedly be of service to those involved in monitoring current pollen levels in the air during allergy studies, but this aspect of the science has not been our main consideration. Neither have we attempted to provide for the needs of palaeoecologists interested in Tertiary and earlier geological strata.

We have attempted to be simple in concept and expression, and to avoid unnecessary jargon and terminology. In subjects such as pollen grain morphology, however, the use of a carefully defined terminology adds simplicity to expression rather than complexity. For those not familiar with morphological terms it is advisable to spend a little time in becoming familiar with these terms (diagrams and definitions are provided) before using the key. Chapter 4 on pollen and spore morphology should help in this respect.

The techniques of pollen and spore extraction are essentially simple. Also the identification of perhaps ninety per cent of the pollen found in post-glacial sediments will present little difficulty. The equipment required for extraction and identification is also fairly straightforward, namely a centrifuge and a microscope. Pollen analysis is, therefore, a technique which is suitable for use in schools as well as higher education establishments. The problems which can be solved by means of the technique are exciting ones, concerned with the history of our vegetation and hence with the evolution of our contemporary environment. It is not surprising, therefore, that the subject has inspired such enthusiasm among its devotees. In this field the opportunity to be involved in truly original research can be placed in the hands of the sixth-former. It is hoped that this book may encourage the development of palynology in schools, especially since it is so well suited to the current project approach in education.

No attempt has been made to describe the results of palynological work in the British Isles except where this affords a useful example of a particular technique or approach to interpretation. The reader is recommended to consult the most readable and lucid account given by Dr. Winifred Pennington in her *History of British Vegetation* (1974). It is hoped that this volume will provide a practical supplement to Dr. Pennington's work.

Many people have assisted in the preparation of this book in a variety of ways and we should like to take this opportunity of acknowledging their help. Dr. D. D. Bartley, Mr. C. Beggs, Mr. P. M. Benoit, Mr. J. Carpenter, Prof. P. A.

Colinvaux, Mr. T. Farrell, Dr. K. Fergusen, Dr. G. Metcalfe, Mr. C. Morris, Dr. K. Rybníček, Mr. D. T. Streeter, Dr. K. V. Symonds, Miss C. Wall and Mrs. W. D. Wall, all contributed pollen material for preparation and photography. Assistance with photography was supplied by Mr. Harry Edge, Miss Laura Neville and Mr. Nicholas Ouzman. We would also pay particular tribute to the work of Mrs. Jennifer Shirriffs, who, under the direction of Dr. S. E. Durno, prepared the scanning electron micrographs of various pollen grains and spores. We are grateful to the Macaulay Institute for Soil Research, Aberdeen, for allowing us to use these photographs. We would also thank the many other authors of previously published material who have kindly granted us permission to reproduce their illustrations, data, etc.

Several students, both postgraduate and undergraduate, have been used as 'guinea pigs' during the evolution of the pollen and spore key. Worthy of especial mention are Dr. S. Handa, Mr. C. Sydes and Miss R. Kalra, all of whom have submitted useful comments and suggestions.

We should like to thank Professor E. A. Bell for providing facilities for this work within the Department of Plant Sciences, King's College, London, and Mrs. V. Lilleberg for her patient and accurate typing of the original manuscript. Our publishers deserve particular thanks for their careful and patient handling of the manuscript.

P.D.M.
J.A.W.

Contents

Chapter 1

The Potentialities of Pollen Analysis

Palynology is the study of pollen grains and spores from higher plants and from cryptogams respectively. Pollen grains and spores are quite different in terms of function, the pollen grain being the container in which is housed the male gametophyte generation of the angiosperm and gymnosperm and the spore being a resting and dispersal phase of the independent gametophyte generation of the cryptogam. Both require dispersal in space, but the pollen grain can only be regarded as having served its function successfully if it arrives at the stigma of a plant of the same species. Cryptogam spores require only to arrive at a site where they can germinate and where the resulting gametophyte plant can establish itself and survive.

Pollen and spores share a common origin, both resulting from meiotic cell division; they are also similar in size and in their need for dispersal. The study of pollen and spores may both therefore be included within the science of palynology.

Palynology is concerned both with the structure and formation of these pollen grains and spores and with their dispersion and preservation under certain environmental conditions. One aspect of palynology is the study of fossil or subfossil pollen grains and spores in both ancient and recent sediments, and it is with this aspect of palynology that this book will be most concerned.

It was in the mid-nineteenth century that fossil pollen grains were first discovered and towards the end of that century they were found in the peat deposits formed since the last glaciation. Scandinavia was the centre of these early investigations where men like C. A. Weber began to extend their work beyond the macroscopic, or large scale, examination of plant remains in peat

deposits to the small-scale microfossils. It was soon realized that such tiny fossils, of the order 20–40 μm, could often be identified with a high degree of precision. This feature, together with their abundance in peat deposits, led to an early recognition of their potentialities as an aid to the reconstruction of the vegetation of past times.

The insight necessary to discern the potentialities of pollen analysis was, however, limited to a few Scandinavian botanists. By the beginning of this century some of these workers had begun to make quantitative counts of different pollen types at different depths in the peat and to express these data in percentage terms. The man who led the field in this quantitative exercise and who must therefore be regarded as the founder of pollen analysis was Lennart von Post, a Swedish botanist. His methods were generally known and used in Scandinavia by the 1920s and began to spread more widely through Europe by the end of that decade.

The techniques have now been applied to suitable sediments in all parts of the world, which has enabled palynologists to begin the task of reconstructing vegetational changes on a global scale.

The value of microfossils

There are several reasons why microfossils, especially pollen and spores, have proved so valuable as indicators of conditions in the past. In the first place they are preserved much more easily than many other parts of plants due to their structural chemistry. The structure and morphology of pollen will be dealt with in detail in Chapter 4; for our present purpose we need merely note that the walls of spores and pollen grains are constructed of a complex material called *sporopollenin*. This material is a polymer of carotenoids and carotenoid esters (Brooks and Shaw, 1968) and under conditions of low microbial activity, especially in waterlogged, acid situations, it is decomposed very slowly. Pollen and spores are therefore preserved in considerable abundance in peat deposits, lake sediments, etc.

The second important feature possessed by these microfossils is their small size and therefore their tendency to be carried some distance from their source, suspended in turbulent air masses. In some respects this is valuable, in others it can prove a problem. The property of transportation means that the pollen grains falling upon a site suitable for their preservation did not all originate in the immediate vicinity. Macroscopic fossils in a deposit, e.g. fruits, seeds, cones, twigs, etc. because of their large size and poor dispersal, are mainly derived from the local flora. This results in an over-representation of wetland species in the macroscopic fossil flora. Pollen grains represent a flora from a wider area of surrounding land.

A disadvantage associated with this property of wide dispersal is that the pollen types which are most buoyant in the air may be carried very considerable distances and hence give a misleading impression when they are found in pollen assemblages many hundreds of kilometres beyond the distributional limits of the species. Differential and long-distance transportation is therefore a factor which must be taken into account when interpreting pollen assemblages in a sediment.

A third valuable feature of the pollen grain and the spore is that its structure and sculpturing may make it a highly recognizable object. This means that identification can often be taken to the species level, though sometimes it is only possible to deduce the genus or the family from which it comes. In this respect some types of macrofossil (e.g. seeds and fruits) may be identified with a higher degree of precision and are therefore more valuable as fossil material.

The fourth useful feature of pollen and spores is the abundance with which they occur in many

sediments. This abundance allows a quantitative recording of the various types to be made, though there are considerable statistical difficulties involved in the treatment of this data and in its final interpretation. These problems will be outlined in Chapter 7.

Because of these features, pollen grains and spores have proved extremely useful to ecologists, palaeoecologists (those concerned with the ecology of the past), medical scientists, criminologists and agriculturalists. Bearing in mind that many problems are associated with the preparation, identification and interpretation of pollen grains, we shall consider now the various uses which have been made of pollen analysis.

Applications of pollen analysis

The applications of pollen analysis include the following:

1. Tracing the history of plant groups and species.
2. Tracing the history of plant communities and hence habitats.
3. Dating deposits.
4. Studying climatic history.
5. Following the course of man's influence upon his environment.
6. Studying the pollen content of the atmosphere and its effects upon human health.
7. Pollen contents of honey (melissopalynology).
8. Criminology.

All these fields of study have in the past made use of palynology.

Tracing the history of plant species

In order to understand the distribution of any species in space, one must study its distribution in time. This is because distributions are always changing, sometimes as a response to changing climate and sometimes in response to other changing ecological factors, including the activities of man. The history of a species will inform us whether it is on the increase or the decrease, whether its range is expanding or contracting. Such information can often lead us to an understanding of what critical factors underlie the limitation of the range of such species.

Many distribution patterns which appear superficially anomalous can be understood by reference to their histories. An example of this is afforded by species with so-called 'disjunct distributions' (Pigott and Walters, 1954). Many species, now rare in Britain, occur as isolated populations in unstable habitats, e.g. marshes, screes, sand dunes, cliffs, etc. Frequently one finds several scarce plants concentrated together in these habitat fragments. Occasionally a study of the past history of such species has revealed that they were once of more widespread occurrence and have become restricted because of changes in the environment. Often such relict species demand open, unshaded habitats, and hence are able to thrive only where tree growth has been prevented by such factors as soil instability.

Many relict populations were abundant during the period at the close of the last glaciation, when forests were not yet established in Britain. Some such species have now been pushed back by competition from more robust species into maritime habitats, e.g. sea buckthorn (*Hippophaë rhamnoides*) and sea plantain (*Plantago maritima*). Other species have become even more fragmented in their distribution patterns as a result of encroaching forest, e.g. perennial knawel (*Scleranthus perennis*), which grows in isolated populations in the Breckland of East Anglia and in two localities in Powys, Wales. Its pollen has been found in an interstadial deposit from the last glaciation at Chelford, Cheshire (Simpson and West, 1958), and in late-glacial lake sediments in Powys and Dyfed (Moore, 1970), which suggests that it also was a component of the Devensian interstadial and late-glacial flora and that it has subsequently been restricted in its distribution by forest spread (Godwin, 1960).

Other members of the late-glacial flora, e.g. broad-leaved plantain (*Plantago major*) and mugwort (*Artemisia vulgaris*) have been suffi-

ciently flexible in their environmental requirements to take advantage of the open habitats created by man in the last 5000 years and have regained something of their former prominence by adopting the rôle of 'weed' species. Only by historical studies can we understand the status of such weed species and differentiate between native and introduced species.

These few examples will serve to demonstrate the importance of pollen analytical studies in understanding the current ecology of our flora. Unfortunately only pollen which is identifiable to the level of species can be very informative in this sphere. Many pollen grains can only be referred to a genus or even a family, and are therefore of limited value. The problems of contamination and long-distance transport of pollen also need to be considered, especially in the interpretation of records of single pollen grains.

Tracing the history of plant communities

When one considers the total pollen assemblage of a sediment sample rather than simply its content of an individual pollen type, one can begin to reconstruct the plant communities of the past. In fact there are many difficulties associated with this exercise (see Chapter 7), especially if one attempts detailed reconstruction of communities. However, a general picture of the major habitats, e.g. deciduous woodland, open scrub, grassland, can be obtained fairly easily from a pollen assemblage. Modern pollen rain studies, however, have shown that one can make errors even in coming to such general conclusions, especially in open habitats (such as tundra) where a high proportion of the total pollen rain may travel considerable distances.

Early pollen analytical studies revealed a sequence of pollen assemblages which succeeded one another following the close of the last glaciation and in this way a broad picture of our changing vegetation was built up (see Godwin, 1975 and Pennington, 1969). The changing habitats represented by these pollen assemblages have been interpreted as responses to changing

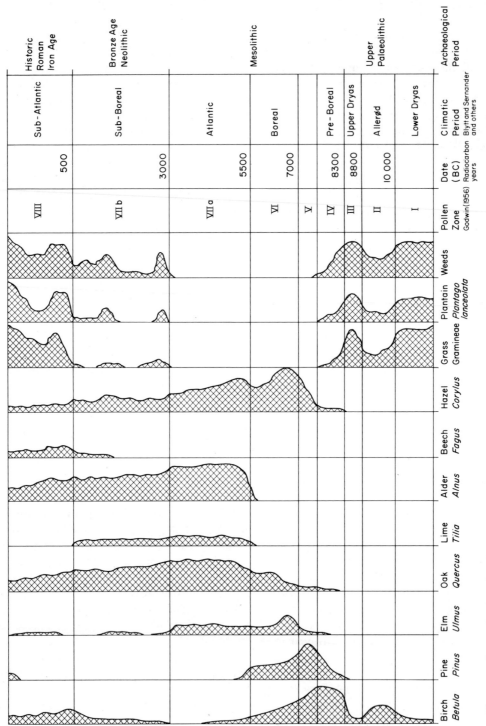

Fig. 1.1 Generalized pollen diagram for the late-Devensian and Flandrian periods in England, showing the pollen curves for selected pollen types expressed as percentage arboreal pollen. The diagram is zoned in the traditional manner (Godwin, 1975) and correlations are shown with radiocarbon dates, climatic periods (Blytt and Sernander) and archaeological cultures.

climates, and on the assumption that such climatic changes were of broad geographical extent and were synchronous over wide areas, such indications of habitat change were used to define 'pollen zones' (Godwin, 1940). These are shown in Fig. 1.1. Since that time some of these assumptions have been questioned, but the broad picture of habitat changes derived from pollen analytical data is still widely accepted.

More detailed reconstruction of plant communities using key species as indicators of phytosociological units has been attempted (e.g. Birks, 1973a; Moore, 1970) but there are many difficulties involved in this process.

Dating

If the major pollen changes which were used for the definition of pollen zones are indeed synchronous over wide areas, then these zones can be dated against some indicator of the passage of time, such as that provided by the decay of carbon-14. Subsequently the pollen content of a deposit could be used to estimate its age within certain rather wide limits. The post-glacial (Flandrian) pollen zones were first systematically dated by Godwin, Walker and Willis (1957) at Scaleby Moss, near Carlisle, Cumberland. These dates are shown in Fig. 1.1. A more recent set of dates has been obtained for zone boundaries at Red Moss, Lancashire by Hibbert, Switsur and West (1971) and in general these agree quite closely with those from Scaleby Moss. When comparisons are made for the whole of northwest Europe, one can detect geographical trends which may reflect the migration routes and speeds of various tree and shrub genera as they migrated through the Continent. Smith and Pilcher (1973) have collated radiocarbon dates from within the British Isles and these show a general lack of synchroneity. Local conditions at certain sites may influence the arrival of different species, particularly factors such as latitude, altitude and exposure (see Moore, 1972).

The same applies to the dating of late-glacial (late-Devensian) pollen zones. Dates were first systematically collated by Godwin and Willis (1959) from a variety of British sites and were shown to be in considerable agreement. The opening of the late-Devensian period is difficult to date, but this is mainly a problem in defining its inception (see Pennington, 1975). The dates of Godwin and Willis are given in Fig. 1.1.

An assumption which has recently been questioned is the validity of the radiocarbon method of dating these deposits. The work of Suess (1970) suggests that the date calculated from radiocarbon values deviates from the true date if the material is more than 2000 years old, and the deviation becomes greater as one goes back in time. However, this is not a very serious problem because Suess has been able to draw up a 'calibration curve' for radiocarbon which allows the conversion of radiocarbon ages into true ages. In fact most research workers are still content to express ages in uncorrected 'radiocarbon years' in full awareness that this does not strictly correlate with the true time scale.

Despite these problems, therefore, pollen data still represent a useful tool in the hands of the archaeologist by means of which he can assign rough dates to deposits. The main limitation to its usefulness is the fact that pollen zones generally occupy fairly extensive time periods, hence dating must be crude.

The reconstruction of climatic history

From pollen analysis it has been possible to reconstruct, at least in part, the history of our flora and this information has been provided with a skeleton of radiocarbon dates. This vegetational history has offered an opportunity for reconstructing the history of our climate, since the plants which produced the pollen have responded to the sequence of climatic changes during the Quaternary period.

Some concepts of changing climates in the Flandrian period had been built up by Blytt and Sernander (see Godwin, 1975) in Scandinavia even prior to the development of palynological techniques. Their work and conclusions were based mainly upon the study of peat strati-graphy. They placed considerable emphasis upon such features as layers of tree stumps in the profiles of mires. The broad conclusions regarding climate which emerged from pollen studies correlated quite well with the findings of Blytt and Sernander, except that the precise climates of their periods have occasionally proved more variable on a geographical scale than they had anticipated. Their climatic periods are inserted in Fig. 1.1. Further work in Denmark led to the subdivision of the late-Devensian period in Europe into a series of cold and mild phases.

Occasionally one can obtain precise knowledge of past climates from palynological data. For example, the presence of *Sedum rosea* in Scandinavia at present is correlated with the 25°C mean annual maximum summer temperature (Dahl, 1951). *Dryas octopetala* is evidently able to withstand slightly higher summer temperatures for its distributional limit coincides with the 27°C isotherm. Such knowledge can be applied to data on the distribution of these species in the past and hence a detailed picture of climatic conditions at any particular period can be developed (Conolly, 1961; Conolly and Dahl, 1970). Unfortunately, a large number of interesting plants cannot be identified to species level on the basis of their pollen (e.g. *Salix* spp), hence the analysis of macroscopic material is often more valuable for this type of work (Dickson, 1970).

Tracing the effect of man on his environment

In recent post-glacial times (Godwin's pollen zones VIIb and VIII) most of the major changes in pollen assemblages can be explained most easily if one considers man as a potential influence upon his environment. Until the work of Iversen in 1941 it had been assumed that the changes which had been observed in pollen proportions, and indeed which had been used as the basis of diagram zonation, were climatically determined. Iversen suggested that certain changes were associated with man, e.g. the periodic increases in non-arboreal pollen to aboreal pollen ratio. Since that time it has become increasingly clear that certain of the major pollen changes which have been used to define zone boundaries could have been caused by human activity, e.g. the zone VIIa/VIIb decline in elm (Troels-Smith, 1954) and the zone VIIb/VIII decline in lime (Turner, 1962).

Intensive palynological work in certain areas has occasionally resulted in the production of a detailed account of the gradual destruction of forest and development of agriculture within that region which can often be correlated closely with historical accounts, e.g. central Wales (Turner, 1964a; Moore and Chater, 1969a) and northern Derbyshire (Hicks, 1971).

Many of the general aspects concerning the interpretation of data relating to human influence upon vegetation in the past have been reviewed by Turner (1970) and Smith (1970). Perhaps one of the greatest uses of pollen analysis in this field is for the prehistoric period, when it provides information on the environment at a period during which we have no documented records.

Aeropalynology

Aeropalynology is the term applied to the study of pollen grains and spores in the atmosphere. A great deal of research has been performed on this subject which has recently been reviewed very thoroughly by Hyde (1969). According to Hyde it was Charles Blackley of Manchester who made the first observations on pollen grains in the atmosphere in 1873. Since that time many scientists have been concerned with this work, but most especially microbiologists, agriculturalists, research workers in allergy and, more recently, palaeoecologists.

The transmission of disease, especially fungal diseases of man and his domesticated animals and plants, by means of spores dispersed within the atmosphere has provided ample incentive for agriculturalists and microbiologists who wish to record and study the aerospora. In addition, the fact that many pollen grains are a causal agent of hay-fever has intensified the monitoring of pollen and spores in the air, leading to the development of a branch of science dealing with the behaviour of these small particles in gaseous suspension (see Gregory, 1961).

Early in the days of palaeopalynology it became evident that these aerial pollen surveys would be of considerable relevance to the interpretation of fossil data. Recent measurements by J. M. Hirst and his colleagues at Rothamsted (Hirst *et al.*, 1967a; 1967b) on the structure and movement of high altitude spore clouds over the continent of Europe have made it even more imperative that detailed work should be undertaken on the behaviour of different pollen types under different meteorological conditions in order that the fossil record can be interpreted correctly.

Pollen rain studies specifically undertaken with this aim in view will be discussed in Chapter 7.

Melissopalynology

The analysis of pollen grains in honey provides information regarding the floral sources which bees are using. This information is often of importance when assessing the quality of the honey (e.g. Lieux, 1972). It is also possible to ascertain the season of honey production since this will correlate with the flowering period of the species represented by their pollen.

Criminology

The fact that pollen grains are recognizable and can aid in the reconstruction of vegetation has led to their use in forensic science. The analysis of soil and mud samples can sometimes provide sufficient information for the determination of their source.

Erdtman (1969) describes an Austrian case history in which a murder was solved eventually by resorting to palynological techniques. A man was arrested and charged with the murder of another man while on a journey along the Danube near Vienna; however no body could be found. Pollen analysis of a soil sample from the arrested man's shoes revealed much pine and alder pollen together with some spores of Tertiary origin. Fortunately only one area was known along the Danube where pine and alder grew together on Tertiary strata, so the suspect was confronted with this fact. He was so shocked at the deduction that he admitted the crime and the precise location where he had hidden the body.

From the foregoing account it can be seen that pollen analysis has considerable potential as a tool for investigating a variety of different problems. Its flexibility is enhanced by the fact that pollen grains are preserved quite well in a variety of sediment types which will be described in the next chapter. The extraction of pollen from such deposits and the preparation of samples for microscopic examination need not be difficult or complex and can be achieved with little need to resort to expensive equipment (see Chapter 3).

Once samples have been prepared, the identification of pollen can be accomplished with the aid of a reasonably good microscope and preferably with the help of a type collection of at least the more important pollen and spores. The principles underlying pollen taxonomy and a key to the identification of types is given in Chapters 4 and 5.

Chapter 2

Deposits Containing Fossil Pollen

Having looked at the various possible applications of pollen analysis, we shall now consider the types of material in which pollen grains may be preserved and from which they may be extracted.

Many types of deposit contain pollen grains, but all of them have certain associated problems which must be borne in mind when analysing and interpreting the pollen assemblages which they contain. We shall consider these problems in this chapter, since such considerations are of importance when deciding what type of deposit to use for a palaeoecological investigation—if, indeed, any choice exists.

Before such a discussion can be undertaken, we need to consider the conditions under which pollen grains and spores are preserved.

Although the coat of pollen grains and spores is resistant to decay, it is by no means 'non-biodegradable'. Some small invertebrate animals, e.g. some springtails (Collembola), appear to ingest pollen grains (Scott and Stojanovich, 1963) and they become severely degraded after passing through the gut of such decomposer organisms. Micro-organisms also are capable of degrading pollen under aerobic conditions at a neutral pH. Because of this the pollen which falls on neutral or basic soils is decomposed fairly rapidly. Even under anaerobic, waterlogged conditions pollen grains are sometimes subject to decay if the pH is high.

Where some degree of corrosion has occurred it is possible that it has acted selectively, in other words some pollen types may have been corroded faster than others. Havinga (1964) has considered this problem and comes to the conclusion that under oxidizing conditions *Betula* pollen is corroded more rapidly than *Quercus* pollen, which in turn is destroyed faster than *Alnus, Corylus* and *Tilia.* Such differential susceptibility is of particular importance when analysing aerobic deposits, such as soils.

The situations in which pollen is likely to be found in a condition suitable for identification and analysis include peats, lake sediments, acid soils and mor humus accumulating at the top of podsol profiles. There are many problems associated with the extraction and interpretation of microfossils from each of these different substrates. Problems of extraction will be dealt with in Chapter 3.

Peats

Peat deposits are accumulations of organic detritus, usually largely of plant origin, which have developed in situations where the rate of production of organic matter by a plant community exceeds the combined rates of plant respiration, herbivore consumption and microbial decomposition. This normally occurs in situations where the decay rate is impaired by waterlogging. The precise circumstances under which peat accumulates and the nature of the peat deposit formed are very varied (Moore and Bellamy, 1974). Table 2.1 gives a summary of the major peat types, classified in terms of the nutrient regime under which they were formed.

The most important feature of peat deposits from the palynological point of view is that they develop in a stratified sequence. Pollen lands upon the surface vegetation and litter, where a certain amount of aerobic microbial decay takes place (see Fig. 2.1). As the vegetation continues its growth, the litter and pollen become buried by further deposits of litter and are thus stratified into horizons within the developing peat profile.

The surface layers of peat (often to a depth of

TABLE 2.1 *Classification of peat types on the basis of conditions of origin.*

Peat type	Origin	Nutrient condition	Examples
Rheotrophic (minerotrophic)	Mire vegetation which receives water both from land drainage and from precipitation	Nutrient rich	Marshes, fens, swamps flushes, carrs, spring mires
Mesotrophic	Intermediate sites where ground water contributes little to the total nutrient capital	Generally nutrient poor	Poor fen, transition mires
Ombrotrophic	Mire vegetation which depends entirely on rainfall for its nutrient input	Nutrient poor	Domed mires, blanket mires

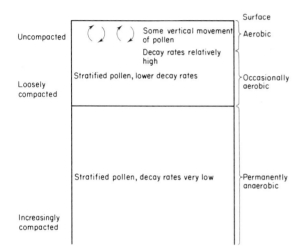

Fig. 2.1 The behaviour of pollen in the peat profile.

about 20 cm in the case of *Sphagnum* peats) is periodically aerated and waterlogged depending upon immediate water availability, and during the periods of aeration there is a development of decomposer activity which causes the breakdown of plant material in the aerated zone. Soluble materials and substances which are easily degraded are removed first and are used as a source of energy by the decomposers (mainly fungi and bacteria). The protoplasmic contents of the pollen grains are lost quickly at this stage, together with the inner wall (intine), leaving the resistant outer wall (exine). Should aeration continue, this component may also become broken down, but pollen exines are one of the last of the peat constituents to suffer decomposition under acidic conditions. As growth proceeds in the vegetation

and as litter continues to accumulate, the forces of capillarity result in a gradual raising of the overall water table. Thus as more litter accumulates, the periods of aeration during dry spells at any given point in the peat profile become fewer and shorter. Decay rates therefore diminish gradually with depth, falling to a minimum when one enters the permanently anaerobic zone (Clymo, 1965). There are some decomposer organisms which continue their activities even in the complete absence of air, e.g. the bacterium *Desulphovibrio* which oxidizes its substrate by using sulphate ions, which become reduced to sulphide. These organisms, together with physical reduction processes, lead to the production of hydrogen sulphide and ferrous sulphide in wet, anaerobic deposits. Decay therefore continues in deeper peats, but at much slower rates than in the surface layers. Again, the pollen exines do not appear to be markedly affected by this anaerobic decay process.

Within a peat deposit one has, therefore, a stratified accumulation of microfossils embedded in a matrix of partly decayed plant detritus. The continued accumulation of peat will cause increased compaction of lower peat layers and sometimes a degree of distortion. This factor, together with the unevenness of the original mire surface, means that the peat shows considerable lateral structural variation when a profile is exposed (Stewart and Durno, 1969). A vertical series of samples will exhibit strict temporal order, inversion within the profile is very unlikely unless severe erosion and redeposition has occurred.

However, it is unwise to assume that peat samples at the same depth but separated by more than a few centimetres laterally are of precisely the same age.

Although vertical movements of pollen up and down the profile becomes most unlikely when the peat is anaerobic and compacted, there is a possibility that vertical movement could occur in the early stages of peat formation. The peat-forming vegetation is often loosely tufted with ample opportunity for vertical water and pollen movement between the stems. As well as the possibility that pollen might move freely in this way, there is also a chance that animals grazing upon the detritus, e.g. springtails, could displace pollen by carrying it in their gut. Undoubtedly a certain amount of this movement does take place. How great such displacement may be and what proportion of the incident pollen is involved is unknown. It is, however, unlikely that such movement will invalidate the basic palynological assumption of vertical (and therefore temporal) stratification because the subsequent compression of the material within which such vertical transport is possible will greatly reduce the effects of pollen displacement. Fresh peat of 10 cm depth can easily become compacted into a vertical space of 1 cm or less lower in the profile and the mixing effects are then lost. Even in the sequential analysis of fresh hummocks of *Sphagnum* peat (within which some vertical movement of pollen is to be expected) the results demonstrate stratified features which suggest that mixing is far from complete (Moore and Chater, 1969a).

In peat deposits developing on unstable sites, e.g. on slopes, there is the possibility of lateral movement of the entire peat mass which could disturb a stratigraphical sequence. This can reach spectacular proportions when mass flow results and a 'bog burst' occurs, but smaller, hardly noticeable movements may occur frequently in peats, such as those described at Muckle Moss, Northumberland (Pearson, 1960).

More radical discontinuities in peat profiles may result from erosion and peat cutting. Many areas of upland peats are suffering from dissection by erosion channels (Tallis, 1964) and the eroded material may be redistributed and become invaded by a new generation of peat-producing plants. Where this has happened there is frequently a hiatus in the peat profile where the *in situ* peat gives way to resorted material. The hiatus is usually detectable by virtue of changes in pollen composition even when it is not noticeable in the stratigraphy. The same applies to old peat cuttings where there has been a regrowth on top of old layers exposed by the cutting. Here a section of the stratigraphy will be missing and the resulting hiatus should be evident both from the peat and the pollen profile. One of the most remarkable pieces of work in which such changes in stratigraphy were recorded was that by Lambert *et al.* (1961) who discovered that the Norfolk Broads in East Anglia had resulted from peat-cutting activities on a vast scale. Discontinuities in peat profiles which have resulted from peat cutting are particularly common in southern and eastern Britain where peatlands have never been abundant and where they have represented an easily obtainable source of energy for local human populations. The pollen analyst must be on his guard when examining peat profiles from this part of Britain in particular.

Even where the peat profile is undisturbed, there are certain other problems associated with peat as a source of material for pollen analysis. The first problem is common to most sediment types, that is the unknown rate of sedimentation. Pollen falling on a mire surface eventually becomes incorporated into the peat, but its concentration within that peat is a function not only of the initial fallout rate, but also of the rate of peat accretion (itself a function of productivity and decomposition rate), and the degree of compaction of the peat. The peat accretion rate will vary with a number of factors, but especially with the nature of the peat-forming vegetation and the wetness. These two factors determine the length of time which any unit of litter (including pollen) will spend in the aerobic surface layers of peat where it is subject to the most rapid rates of decay. Once it is engulfed by the permanent water table the decay rate will fall off.

The degree of compaction of peat will depend upon its physical structure and composition and its depth in the profile. A woody peat will be less liable to compaction than one formed from sedge leaves, and the greater its depth the greater the weight of peat above it.

The outcome of these two variables is of considerable importance for palynology. Although the stratigraphical sequence of peats reflects temporal development, with the deepest peat layers being the oldest, the relationship between depth and age is not a linear one. The relationship is an extremely complex one because of the factors mentioned above. A 10 cm section of peat can represent a time-span of ten years, a hundred years, a thousand years or more, depending upon the outcome of peat growth rates and compaction. Thus, if the total fallout of pollen from the atmosphere were constant over a long period of time, the density of pollen in the peat deposit would still vary with these other factors. This means that when one analyses a deposit for pollen, one cannot meaningfully express the data in absolute terms (i.e. number of grains of species X per square centimetre surface per annum) unless the relationship between depth and age is known. In most cases this relationship is not known, hence data has to be expressed in proportional terms (i.e. number of grains of species X as a percentage of the total pollen, or tree pollen, in the sample counted). This problem and its implications will be discussed at greater length in Chapter 6. The problem is one common to all deposits which accumulate at variable rates, i.e. peats and lake sediments.

A second problem associated with peats is that the peat-producing vegetation itself may give rise to a large pollen input of a local nature. In pollen analysis one is normally interested in regional rather than local vegetational change hence this input represents a dilution of the information we are trying to extract, a kind of background 'noise' which may obscure the reception of information. The extent and the importance of this noise will vary with the peat-forming vegetation. Moss peats, e.g. *Sphagnum* peats, have a very small input of immediately local pollen, and the spores of *Sphagnum* itself are easily recognized, hence can be separated out from the regional pollen rain. Reedswamps produce larger quantities of pollen, particularly grass and sedge pollen, which cannot always be distinguished from the regional grass pollen input. For this reason the pollen analysis of peats should always be accompanied by an attempt to determine the composition of the contemporary peat-forming vegetation from macrofossil remains in the peat profile.

An excellent review of the literature concerned with the identification of macrofossils within Quaternary deposits is given by Dickson (1970) and an extensive summary of the occurrence of such macrofossils in the deposits which have been examined is given by Godwin (1975). Bryophyte material is specifically dealt with in the monograph by Dickson (1973).

Lake sediments

Most lake sediments contain a very large proportion of material which has been derived from outside the lake basin (*allochthonous* material) whereas the bulk of peats develop from vegetation *in situ* (*autochthonous* material). The lake sediment consists generally of inorganic and organic allochthonous detritus, organic material from organisms growing within the lake and an input of pollen direct from the air, from surrounding regions via drainage water and secondarily transported pollen from microfossil-containing deposits which have been eroded by drainage waters (see Fig. 2.2). The proportions in which these various constituents occur vary with the nature of the drainage basin and the organic productivity of the lake. Table 2.2 gives a summary of the major types of lake sediment classified according to the nutrient status and hence the productivity of the lake in which they formed.

The situation is made more complex, however, by the fact that organic material need not all be produced in the lake itself. Just as pollen may be

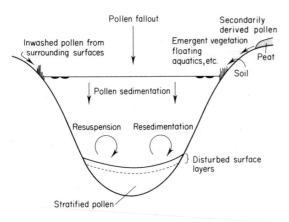

Fig. 2.2   The behaviour of pollen in lakes.

derived from the erosion of pollen-containing deposits in the lake catchment area, so may organic carbon. Plant detritus, peat and organic matter from soils may be eroded and transported into a lake. Hence the carbon content or organic matter content of a lake sediment is not always a good indicator of its contemporary productivity.

Like peat deposits, lake sediments develop in a stratified sequence, hence depth is related to age, though the relationship is not necessarily a linear one. As Table 2.2 indicates, different types of lake sediments accumulate at different rates. Lake sediments do not suffer from all the disadvantages found in peat deposits. They do not experience the same downward movement of

water through the profile which occurs in peat deposits, hence there is far less likelihood of downward movement of pollen through the profile. On the other hand, there are detritus-feeding animals within the upper layers of sediments which could result in mixing. One further feature which results in mixing the surface layers of sediments is water turbulence. This has been examined in detail by Davis (1968) in the United States. She measured sedimentation rates in lakes by two methods. In one she took cores from the surface layers of the lake sediments and analysed these for pollen. The types of pollen present changed very abruptly at a depth of about 25 cm, trees giving way to grasses and weeds. This corresponded to forest clearance and land settlement in the area (Michigan, USA) in 1830. From this it was possible to work out sedimentation rate since that date and also (from measurements of pollen density within the sediment) the rate of pollen sedimentation per unit surface area. The second method of measurement was to use 'sediment traps'. These were open bottles which were suspended, open neck upwards, two metres above the bottom of the lakes and in which sedimentation rates and pollen accumulation rates were measured. The rate of pollen accumulation in these sediment traps was between two and four times faster than indicated by the sediment cores. Detailed

TABLE 2.2   *Classification of lake sediment types.*

Lake type	Nutrient availability	Sediment composition	Rate of accumulation
Eutrophic	High, leading to high productivity	Large proportion of organic material termed *gyttja* or *neckron mud*. Precise nature varies with producer organisms involved, e.g. algae, broad-leaved aquatics. Highly calcareous deposits are termed *marl*	Generally rapid
Oligotrophic	Low, therefore low productivity	Largely minerogenic, i.e. low organic matter content	Slow, depending on erosion rates in the catchment
Dystrophic	Low nutrient status and low mineral input, therefore low productivity	Colloidal precipitates called *gel-mud* or *dy*, rich in pollen but poor in other types of organic detritus	Slow

analyses of the pollen input into these traps revealed that pollen accumulated most rapidly in the periods October–November and April–June. In the case of the October–November and the April periods, the vast bulk of the pollen sedimenting into traps was of species which were not flowering at that time of year. This pollen must have been resuspended from the surface sediments and was being deposited for a second time. Only during May and June was the pollen deposition made up largely of newly arrived grains from the flowering plants of surrounding areas.

The periods during which pollen resuspension and redeposition occurred corresponded with times when the lake was not frozen, but was not thermally stratified and stable. During the summer months the surface waters become warm and lie in a stable fashion above the cooler waters below. In this condition there is little or no disturbance of surface sediments. When this stratification is broken down by falling temperatures and rising winds in the autumn, there is a resuspension of these sediments. With the freezing of the lake some stability returns to the sediments, but this is lost again when the lake thaws in spring until eventually the summer stability sets in.

An additional complication is provided by the fact that erosion of the surface sediments tends to be greatest in shallow water near the lake margin, but redeposition takes place evenly over the whole basin. This results in the actual accretion rate of sediment being lower in shallow water than in deep water.

There are two major implications of this study for standard pollen analytical work. In the first place, the lateral movement of pollen which occurs in lake waters will tend to reduce the variation one might expect between one core and another in terms of proportional pollen assemblages (rather than absolute, quantitative ones). This means that one has greater justification for taking a single core as representative of the entire basin. The second implication is that each year's pollen input is mixed with that of previous years

prior to deposition. This means that a sediment sample from a given vertical position in a core represents not the current year's pollen input, but the average of several years' input. This has the effect of reducing variation in pollen contents between adjacent samples or 'smoothing' the changing pollen proportions.

Thus these redepositional phenomena within lake basins can probably be regarded as an advantageous feature of lake sediments when considered as material for pollen analysis.

Other types of sediment redistribution within lakes may be less advantageous. Streams entering a lake bring with them a load of material eroded from the surrounding watershed and deposition of this material, especially the larger particles will occur near the point of entry. Scouring by entry and exit streams, on the other hand, may remove lake sediment, producing a hiatus in the stratigraphic profile similar to that caused by erosion in peats.

Problems such as these can be avoided by sampling sediments which show no evidence of disturbance of this type. These are likely to occur in the deepest parts of the lake basin. The problem of pollen rebedded from surrounding sediments, however, is unavoidable because sedimentation of such pollen is likely to occur throughout the lake basin. The types of pollen and spores which are brought into the basin will depend upon the type and age of material being eroded by incoming streams. Sometimes the eroded sediments are of considerably greater antiquity than the contemporary sediments and the fossils they contain are of a quite different nature. The rebedded material is then not difficult to identify. Occasionally one can detect the presence of rebedded pollen by this method, but one cannot be certain how great the rebedded component is. For example, Davis (1961) found evidence of pre-Quaternary tree genera in late-glacial sediments in Massachusetts which must have been derived from erosion of Tertiary deposits twenty-five miles away, presumably carried in glacial meltwater. In this case it was impossible to be sure how much material had been rebedded because the composition of the Tertiary and Quaternary floras overlapped in certain respects. Davis considered the entire data unreliable because of this failure to separate totally the re-bedded pollen component.

Cushing (1964) experienced similar problems in late-glacial (late-Wisconsin) sediments from Minnesota. However, he attempted a separation of the rebedded component on the basis of the types of deterioration of the pollen exine which had occurred as a result of the differences in history. This technique was developed and placed upon an objective footing (Cushing, 1967a) in which six classes of exine deterioration were recognized: (i) corroded grains, exines etched and pitted; (ii) degraded grains, structural elements of the exine rearranged and apparently fused; (iii) crumpled grains, folded with the exine collapsed and thinned; (iv) crumpled grains with a normal exine; (v) broken grains, with the exine ruptured; (vi) well-preserved grains. Cushing found that different types of deterioration were associated with different sediments in his Minnesota lakes, suggesting that this could provide a means of identifying rebedded pollen precisely. Birks (1970), working on similar-aged sediments in the Isle of Skye, Scotland, was able to go further than this and separate the rebedded component into 'secondary' pollen, i.e. pollen derived from eroding deposits of greater antiquity, and 'inwashed' pollen, which was contemporary pollen washed into the basin from the surface of local vegetation, e.g. mosses. Secondarily derived pollen was associated with silts and contained a high proportion of broken and degraded grains (*sensu* Cushing) while inwashed pollen tended to be corroded by aerial oxidation and microbial decay prior to its incorporation in the lake sediment.

The critical examination of the state of deterioration of pollen grains may well offer an opportunity for the full separation and analysis of the rebedded component in lake sediments. More recently Phillips (1972) has identified pollen and spores from older sediments on the basis of their different fluorescence colour and intensity when examined by fluorescence microscopy.

As in the case of peats, there is a contribution to the pollen input of lakes from plants growing locally, in this case aquatic plants. In peats this component is difficult to distinguish from the regional component because local plants may be of the same pollen type, e.g. Gramineae, Cyperaceae, Ericaceae, etc. In lakes, however, the aquatic plants are usually of a quite different pollen type from regional components, e.g. *Potamogeton*, *Myriophyllum*, *Nymphaea*, etc., hence the separation of the immediately local component is considerably easier than with peats. There is, however, the possibility of mire vegetation around the lake which will contribute to the pollen input, e.g. reedbeds contribute Gramineae-type pollen and birch carrs contribute *Betula*. As with peats one must rely upon detailed stratigraphical examination of the entire basin to elucidate these effects.

The different rates of sedimentation under different environmental conditions in lakes have already been mentioned. As with peats, this means that unless one knows the rate of sedimentation and the density of pollen within the sediment one cannot express pollen frequency on an absolute basis (number of grains/cm^2/annum). Usually it has to be expressed in relative terms, as a percentage of tree pollen, total pollen or some other appropriate proportion.

Soils

In the early days of pollen analysis, soils received very little attention. It was generally assumed that pollen would be broken down by microbial activity in aerated soils. Even if it were to survive it was expected that the activity of soil organisms would serve to mix it with the inorganic substrate so that any stratification would be lost. Erdtman (1943) then demonstrated that certain soils contain very considerable quantities of pollen, up to half a million per gramme, and Dimbleby (1957) gave estimates of up to 1·5 million grains per gramme of dry soil. This level of preservation, however, is achieved only in soils of pH less than 5. Above a pH level of 6, there is virtually no preservation of pollen. Whether the soil is a podsol or a brown earth, or whether it is light or heavy in texture, does not appear to affect the level of preservation. Temperature also appears to have little effect since pollen is even preserved in acid tropical soils.

Having established that pollen can survive in certain soils, one is still faced with the problem of pollen movement, both downwash and mixing due to the effects of soil animals. Early experiments by Mothes, Arnoldt and Redmann (1937) indicated that the simple washing of pollen through sandy soils could be quite a rapid process, but in natural conditions it is likely that such movement is made slower by the embedding of pollen into humus material which moves more slowly. Soil compaction also appears to slow down the rate of movement. Old root channels, on the other hand, contain high concentrations of pollen, presumably moving down the channels.

Despite this movement, analysis of soil profiles has frequently revealed a distinct stratification of pollen (e.g. Dimbleby, 1957; 1962). The stratification, however, should not be regarded as a temporal one, neither can a pollen assemblage at a particular level be regarded as representative of a contemporary flora. Different pollen types move down the profile at different rates, which means that a single horizon may contain pollen of a variety of ages. The presence of an iron pan in the soil profile results in the accumulation of pollen in the region above this impedance to further movement. This layer represents the base of the stratified zone.

These various problems combine to make the interpretation of pollen stratification within soils extremely complex (see Dimbleby, 1957 and 1961).

Sometimes soils become 'fossilized' as a result of burial, either by natural means, e.g. aeolian (wind blown) sands under periglacial conditions, or by man, e.g. the erection of burial-mounds over contemporary soils (Dimbleby and Speight, 1969). Often the analysis of such soils can reveal

much conerning the type of vegetation in existence prior to the burial. By this means Dimbleby (1952) was able to show that the North York Moors carried deciduous woodland before the clearance and erection of barrows in Bronze Age times. When dealing with buried soils of this type, the stratification of pollen within the soil is of less interest than the overall pollen composition of the soil. Under such circumstances the difficulties associated with pollen movement are less important.

Mor humus

Under certain vegetation types, e.g. pine and heather, litter is incorporated into the mineral soil only very slowly. This appears to be associated with a high polyphenol content in the litter and a low rate of breakdown by microorganisms (Coulson *et al.*, 1960; Davies *et al.*, 1964). Such litter accumulates on the surface of a soil in a stratified manner similar to that of peat deposits. Indeed the initiation of certain types of blanket peat may be preceded by this process (Taylor and Smith, 1972). This means that such deposits can be subjected to pollen analysis in the same way as peats and the problems associated with interpretation will be very similar to those found in peat deposits.

Generally such mor humus layers are rather thin, hence the scope for reconstructing vegetational change is limited, however deposits of up to 70 cm have been recorded and analysed in Denmark (Iversen, 1964). Under such circumstances a considerable period of vegetational development may be recorded.

It can be seen that a large number of different types of deposit can be analysed palynologically. In terms of potential as a record of vegetational change, each has certain advantages and disadvantages over the others. These factors must be considered carefully in the choice of a site for palynological investigation and in the interpretation of the data obtained from such an investigation.

Chapter 3

The Collection and Treatment of Samples

The materials discussed in Chapter 2 (peats, lake sediments, etc.) have in common the characteristic of being stratified in a sequence which bears some relationship to time. Since these deposits occur in a chronologically stratified sequence it is important that they should be collected in such a manner that this sequence will not be disturbed.

Peat and sediment samplers

A number of different instruments have been developed for the recovery of cores with a minimum of disturbance and three of these are shown in Fig. 3.1. As an alternative to sampling with these instruments, it is sometimes possible to cut profiles of deposits and to remove a cut sequence in the form of an intact monolith (e.g. in the case of shallow peats).

Hiller borer

For most fibrous peat deposits the traditional sampling instrument has been the Hiller borer (Fig. 3.1a). This instrument does have certain disadvantages.

1. The projecting flange of the rotating chamber, as well as the auger head, are liable to catch resistant root and shoot material in the surface layers and carry them down to some depth, thus resulting in contamination. This can be overcome in part by replacing the auger head with a steel cone, but the problem of the flange remains.
2. The auger head disturbs the deposit as it penetrates so that the material ultimately sampled may have been churned.
3. The auger also disturbs material to a depth of 10–20 cm below the zone of sampling. This

can be overcome by using two adjacent holes for the recovery of alternate half-metre cores; this can be done only where lateral variation in stratigraphy is small (see Deevey and Potzger, 1951).

4. In the conventional Hiller, intact cores cannot be removed and samples must be taken in the field. This is a serious disadvantage for a number of reasons. In the first place it is impossible

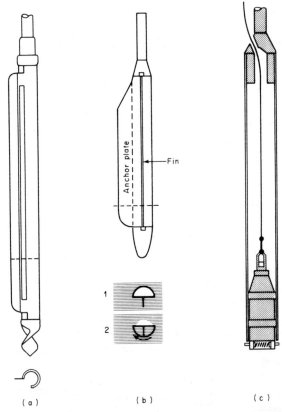

Fig. 3.1   Three of the most important types of peat and sediment sampler (after West, 1968): (a) Hiller borer; (b) Russian (Jowsey) borer; (c) Livingstone (piston) corer.

to maintain an adequate degree of hygiene in the field for pollen sampling. Surface contamination is inevitable, especially if cores are taken during spring or summer (when atmospheric pollen is abundant), or under difficult weather conditions when operators may be working at sub-optimal efficiency. In the second place, it may subsequently become desirable to sample at closer intervals at certain points in the core. If the sampling interval must be determined in the field then this is impossible. In regions of rapid peat accumulation or slow vegetational change a sampling interval of 10 cm may be adequate, but if sedimentation is slow, or past pollen rain changes rapid, then it may even be desirable to take contiguous samples of material. This option is not open for material gathered with a conventional Hiller borer.

Thirdly, the reconstruction of a detailed stratigraphic sequence is essential for the interpretation of any pollen diagram and such a reconstruction is impossible without a complete core.

A modified version of the Hiller borer (Thomas, 1964) overcomes this difficulty. Its auger head is removable and the rotating inner chamber can then be slid out. It is possible to use liners of plastic or zinc within the inner chamber which then permit the removal of intact cores. These can then be wrapped in polythene and sealed, labelled with site, depth and which way they are orientated vertically, and then taken to the laboratory for analysis.

Although this instrument is a great improvement on the conventional Hiller borer, it still has contamination problems.

5. A final problem with the Hiller borer is that of cleaning it adequately. Again this process is facilitated in the modified Hiller in that it can be dismantled.

Russian borer

An alternative instrument for sampling peats and well consolidated limnic sediments has been described by Jowsey (1966) and is often referred to as the 'Russian borer' because of its design origin (Fig. 3.1b). This differs from the Hiller in that it cannot be rotated during descent and in that it samples only a semicylindrical core of sediment.

It has several advantages over the Hiller:

1. It is not as liable to the problem of carrying down superficial material because of the absence of an auger head in addition to which the flange of the fin slopes gradually and does not have projections in which roots can be trapped.
2. It does not disturb the deposit which is to be sampled. In descending vertically in a closed position it slides past the material which will be sampled, hence disturbance is minimized.
3. The core which is retrieved is fully exposed on the fin (after rotating the chamber). This core can easily be slid into a container and taken to the laboratory intact.
4. The process of cleaning is extremely easy in that it involves simply the wiping over of the fin surface.
5. It is very much faster to operate than the modified Hiller in that no dismantling is necessary.
6. It has a considerable advantage in the study of detailed stratigraphy in that a large surface area of undisturbed sediment is fully exposed on the fin, which considerably facilitates the recognition of changes in humification, colour, banding, etc. in the deposit.

The Russian borer has one serious disadvantage when compared with the Hiller, and that is the difficulty involved in using the instrument in stiff sediments. The borer can only be pushed vertically into the deposit and if stiff clays or wood are encountered they may be difficult to penetrate. Field workers engaged in the boring of peat mires are well advised to be equipped with both types of head.

Piston sampler

Work in well-decomposed, sloppy peats and lake sediments is best performed with the use of some type of piston sampler (Fig. 3.1c). Several designs are available, perhaps the best known being the Livingstone modification of the Dachnowsky sampler (Rowley and Dahl, 1956). This consists of a movable piston in an open-ended cylinder. Once at the required depth, the piston is held rigid while the cylinder is pushed further into the deposit. The sample is then withdrawn with the piston at the top of the core.

In most corers of this type samples of 0·5 metre or 1 metre in length are taken in sequence and casing can be used to prevent slumping and to relocate the sampling hole, particularly if this is underwater. Mackereth (1958) has designed a modification by which cores of a length corresponding to the depth of the entire sediment can be obtained in a single, underwater coring. The core is recovered under pressure.

Difficulties in the use of this type of instrument include the problem of penetrating very stiff sediments. Also the sampling of superficial lake deposits is very difficult. Some attempts have been made to design a sampler which will freeze the top half metre of a lake sediment *in situ* before retraction (e.g. Shapiro, 1958). This technique allows the loosely compacted, superficial sediments to be sampled without disturbing their stratigraphical sequence.

The sampling site

All these instruments provide a means of retrieving a core of sediment from which macrofossils and microfossils can be recovered. Because accurate pollen analysis is a time-consuming

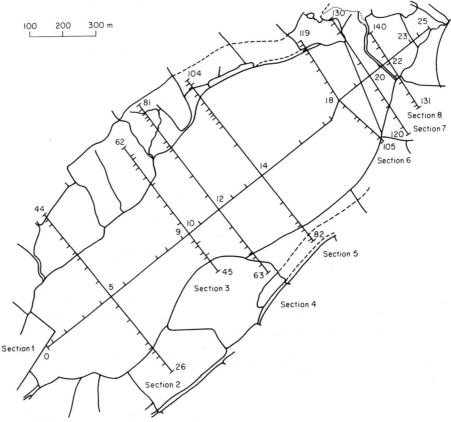

Fig. 3.2   Outline map of a domed mire in Radnorshire, Wales (Rhosgoch Common), showing the system used by Bartley (1960) for investigating its stratigraphy. Two stratigraphic sections are shown in Fig. 3.3 and Fig. 3.4.

process, most workers are contented with one complete core from any site for palynological study. However, it is desirable that the overall sediment depth and stratigraphy should be examined over a wide area before the selection of a site for pollen coring. Not only does this allow the final site to be chosen with precision, it also

degree to which the sequence changes between bore holes. Where the deposit is uniform, the bore holes can be widely spaced, but where adjacent boring produces different sequences, intermediate borings should be taken until the overall stratigraphic relationships of sediment types can be adduced. Figs. 3.3 and 3.4 show two

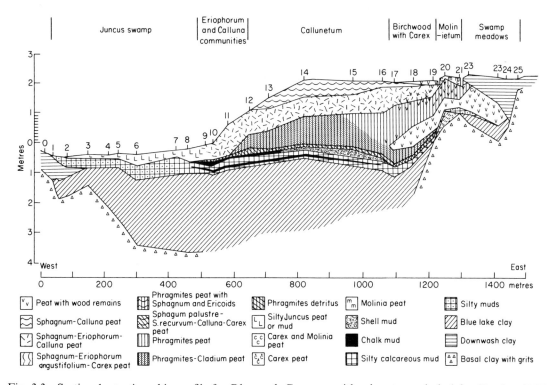

Fig. 3.3   Section 1: stratigraphic profile for Rhosgoch Common with a key to symbols (after Bartley, 1960).

provides information which may be of considerable relevance to the interpretation of the pollen diagram.

Most sites can be studied effectively by means of the reconstruction of a series of stratigraphic profiles built up from the data supplied by a number of bore holes. These bore holes should be arranged along lines at right angles to one another, as shown in Fig. 3.2, which shows the arrangement of bore holes selected for the examination of a mire in Wales (Bartley, 1960). The spacing of bore holes depends upon the complexity of the stratigraphy, particularly the

of the stratigraphic profiles from the site shown in Fig. 3.2. The production of such stratigraphic profiles requires that the mire surface should be surveyed accurately to assess the vertical height of each sampling point.

Having completed such a preliminary survey one is in a far better position to select a suitable location for taking the core destined to be subjected to pollen analysis. The criteria for selecting this location are as follows:

1. Depth of sediment. Normally one wishes to collect the most complete sequence of deposits,

which frequently, though not always, means the deepest deposit. Even in a situation where sedimentation or peat accumulation began uniformly over a whole basin there are advantages to be gained in choosing deep deposits; one of these is that in such deposits a greater depth of material will have accumulated per unit time, which allows a higher degree of

such as birch, willow, alder, etc., have been growing locally throughout the course of mire development. Cores through such regions, or areas adjacent to them, should be avoided since one can expect over-representation of these taxa in the pollen spectra. Usually the avoidance of local over-representation means the selection of central areas of mire sites.

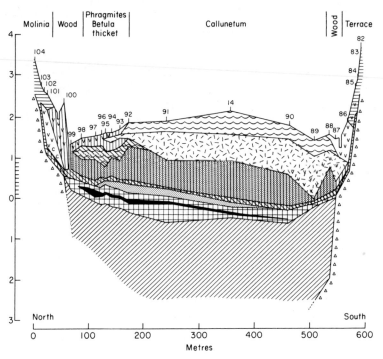

Fig. 3.4   Section 5: stratigraphic profile for Rhosgoch Common (after Bartley, 1960). See Fig. 3.3 for key to symbols.

resolution in the pollen sequence in that samples can be spaced more closely and hence a more detailed picture of vegetational change can be built up.

2. Avoidance of local pollen influences. When one is aware of the stratigraphical development of an entire site, it beomes possible to attempt to avoid collecting pollen cores from situations in which strong over-representation of certain local types can be expected. For example, many domed mire sites have peripheral regions (lagg) in which tree species,

3. Avoidance of areas where sediment disturbance is suspected. Regions of mires and lakes which have been eroded or disturbed by runoff or stream entry respectively, can sometimes be detected by a consideration of stratigraphic profiles. These areas should obviously be avoided in the selection of a pollen sampling site.

As well as assisting in the choice of pollen coring site, a knowledge of the full stratigraphic development of an area is of considerable value in the interpretation of pollen diagrams. Often certain elements in a pollen assemblage can be referred to contemporary plant communities of peripheral parts of the mire or lake, whose existence may have gone unsuspected if only a single core had been taken.

Transport and storage of cores

As has been stated previously, wherever possible, sampling of peat cores in the field should be avoided. It is far preferable to take intact cores to the laboratory where they can be sampled far more accurately and comfortably. Cores are most effectively collected in rigid zinc or, better, plastic casings (longitudinally split plastic drainpipes are ideal). They must be fully and indelibly labelled, including an indication of site, boring location, depth and which end is uppermost in the stratigraphic sequence. They can be wrapped and sealed in clean polythene and transported in robust boxes. Ex-army ammunition boxes are very useful since they are durable, have handles at each end and they can be obtained in the appropriate size (i.e. with an internal length of 50 cm).

It is often necessary to store cores or monoliths for considerable periods of time before they can be analysed. This is possible if care is taken to discourage any rise in decomposition rate. This means that cores should be effectively sealed (e.g. with polythene) to prevent water loss and the development of aerobic conditions which encourages microbial activity. It is also sensible to keep the material at low temperature, both because this reduces the tendency for evaporation to take place and also because it lowers the metabolic rates of any microbes which may be active. Freezing of cores is not generally to be recommended since this causes an increase in volume which may influence subsequent sampling positions. It may also distort sequences in rigidly held cores. Storage at about 4°C has been found most effective, and at this temperature cores may be kept undamaged for many years.

Sampling for pollen extraction

In most studies it is not possible to use the entire core systematically for pollen extracts. Instead subsamples are taken at selected depth intervals along the core. The intervals at which samples should be taken depends upon a variety of factors, such as the rate of growth of the deposit, the rate of change in pollen assemblages with depth, the degree of precision with which it is desired to detect such changes and the time available to the investigator for the project.

A commonly adopted sampling scheme is to begin by sampling at fairly wide intervals, such as 20 cm or 10 cm and then to fill in gaps where appropriate. This approach allows the construction of a 'skeleton pollen diagram' based upon wide sampling intervals which allows the palynologist to detect regions of the core which exhibit the most rapid changes in contemporary vegetation or which are of particular interest for some other reason. These can be investigated in more detail by the analysis of intermediate samples. This approach is economical on time and effort, but is possible only when entire cores or monoliths are available in the laboratory.

There is one minor disadvantage with the use of 20, 10 and 5 cm sampling intervals, that is the difficulty of sampling at points intermediate to 5 cm samples. The recommended quantity of material for pollen extraction is approximately one cubic centimetre and this is normally extracted from the core in such a way that the sample occupies one centimetre depth of the core. It is this fact which makes close sampling awkward if one has begun sampling on a decimal system. A simple alternative is to begin 'skeleton' sampling at 16 cm intervals, which allows closer interval sampling to be used at 8 cm, 4 cm, 2 cm or even 1 cm if necessary. This means that if a high degree of resolution is required from a diagram one can insert intermediate samples at closer and closer intervals until eventually contiguous one centimetre samples of peat are being analysed.

Cutting the peat sample from the cores should

be preceded by a careful cleaning of the core surface. This should involve the cutting away of superficial material, which is that most likely to be contaminated. This cleaning should be undertaken with a clean, sharp scalpel and all cleaning cuts should be made in a direction horizontal to the axis of the core. In other words, movement of the scalpel should always be sideways to avoid contamination of any given level of sediment with material from a position above or below its own.

Unless sampling is to begin at close intervals, there is little merit in cleaning an entire core initially. It is better to clean only the area around the position of sampling, since rewrapping and storing cores is bound to cause further superficial contamination.

The 1 cm^3 sample can be cut out with a scalpel or spatula. In all but absolute methods (see later) there is no need for a high degree of precision in determining the size of the sample, since the pollen content will be expressed in relative percentage terms and will thus not be affected by sample size. Precision is necessary if pollen density is to be used as an indication of sedimentation rates (see Conway, 1947; Moore and Chater, 1969b).

All of these, and subsequent, treatments should be carried out in an atmosphere which is kept as far as possible clear of pollen contamination. Simple precautions are to avoid bringing sources of pollen into a preparation laboratory, e.g. herbarium specimens and potted plants, to avoid making up type specimens and test samples in the same equipment or the same laboratories, and to try to avoid periods of high aerial pollen for preparing specimens. On this last point, some workers prefer to complete all field work and slide preparation during the winter. If preparations are made in summer, then the afternoon should be avoided since this is the time of greatest pollen release. Contamination of samples by aerial pollen can be recognized if it occurs at a late stage in preparation since such grains may not be stained and will still contain their cytoplasmic contents. However, contaminant modern grains from an early stage in prepar-

ation, such as those arriving while cutting a sample from a core, cannot be distinguished from subfossil pollen.

Pollen extraction

The preparation of a sample suitable for pollen counting from a Quaternary deposit does not strictly involve *extraction*, but rather *concentration*. The techniques involved in the process are aimed largely at the disintegration and dissolution or otherwise removal of the non-pollen matrix in the sediment. The various ways in which this is achieved are based upon the small size of pollen grains and spores, coupled with their extreme resistance to many caustic chemicals.

The complexity of the pollen concentration process depends upon the original concentration of grains and the nature of the matrix. In most ombrotrophic peats (see Chapter 2) the matrix is entirely organic. In rheotrophic peats and most lacustrine sediments (and also in soils) there are varying quantities of mineral material, either siliceous or calcareous. If the concentration of mineral material is at all high, then it becomes necessary to remove it.

The ultimate object of this concentration exercise is that pollen-rich samples shall be produced, mounted on microscope slides, which shall permit: (1) the accurate identification of as many as possible of the grains encountered; and (2) the counting of an adequate sample of pollen grains to provide a representation of the total population (see Chapter 6). Therefore elaborate and time-consuming preparation techniques should be avoided if simple techniques can be used to produce samples which are sufficiently rich in pollen to count and in which problems of obscuration by matrix material are not too recurrent. If pollen concentration is low or if identification is impeded by matrix material, then further processing will prove necessary. If countable, clear samples can be produced by simple processing (e.g. KOH digestion and acetolysis), then further treatment is liable to prove a waste of time.

Pollen concentration techniques will now be presented, beginning with the simplest of processes and proceeding to more elaborate techniques which will prove necessary in difficult sediments or those with low pollen contents.

1. *Potassium hydroxide digestion*

This method is suitable as a sole treatment only for organic sediments with no or very low mineral contents and a relatively high initial concentration of pollen. In all other sediments this treatment represents simply a first step in a more extensive programme.

(a) Place 1 cm³ of a sample in a boiling tube and add 10 ml of 10% KOH. Place for 20 minutes in a boiling water bath, stirring to break up the material with a glass rod. Ensure that the concentration of KOH does not rise above 10% by the occasional addition of distilled water. Prolonged boiling is unlikely to damage pollen as long as the concentration does not rise too high.

(b) Sieve the material through a 100 μm aperture diameter sieve and wash through with distilled water. Pollen will pass through the sieve together with other microfossils (e.g. fungal spores), fine fragments of plant and animal debris and humic materials liberated by KOH treatment. Left on the sieve will be the larger fragments of debris which may be discarded or, better, stored for subsequent examination for macroscopic subfossils which may give information relating to local flora. Material passing through the sieve should be collected in a polythene centrifuge tube.

(c) Centrifuge for 3 minutes at about 3000 r.p.m. This results in a pellet of pollen-containing debris collecting at the bottom of the tube. It is preferable that a swing-out head centrifuge be used rather than the solid head type with inclined tubes. The latter results in a smear of deposit up one side of the tube.

(d) Decant off the liquid in the tube. This consists largely of organic soil colloids and can be discarded.

(e) Resuspend the pellet in distilled water, agitating either with a stirring rod or a mechanical stirrer. Occasionally it is difficult to break up the material into a fine suspension and 'clumping' occurs. Should this happen it can usually be overcome by adding a few drops of 5% sodium lauryl sulphate solution or any similar detergent.

(f) Centrifuge as before, then repeat the washing process in distilled water as in (d) to (f).

(g) After final centrifugation, add two drops of aqueous safranine to the tube together with about 4 ml of distilled water and agitate thoroughly. Then top up with distilled water and centrifuge once more.

(h) After decanting, a clean, stained pellet of debris will remain which should be enriched in its pollen content. Molten glycerol jelly may now be added to the pellet and the stained material suspended evenly within it by thorough mixing. The precise quantity of jelly which should be added depends upon the quantity of material surviving the treatment. Over-dilution at this stage leads to difficulties when counting, while under-dilution can produce a medium in which microfossils are so densely crowded as to obscure one another. Usually between 1·0 and 1·5 ml produces a workable suspension if 1 cm³ of peat is used initially.

(i) The jelly suspension should be kept warm in a water bath and microscope slides should be warmed on a slide-drying hotplate or a photographic warming plate. Some people prefer to spread the suspension over the whole area to be covered by the coverslip whereas others make a central smear which spreads when the coverslip is placed upon it. if the latter method is used, then one must take into consideration the differential movement of pollen in the spreading jelly. Light grains tend to be more mobile and hence end up on the edges of the coverslip (Brooks and Thomas, 1967). This can be overcome by using traverses of the microscope along the axis of pollen movement when counting.

At least two slides should be made up for each level to reduce sampling errors (see Chapter 6).

(j) It should not be necessary to seal slides if they are to be counted within a reasonable period, e.g. a year or so. If it is desired to keep them longer than this then it may be necessary to seal them with clear cellulose solution in acetone. They can be stored (after labelling with site and depth!) either vertically or horizontally in a cool place.

This process of concentrating pollen is rapid, simple and effective for the majority of peat types which have no mineral (i.e. silica or calcium carbonate) content and which are fairly rich in pollen in the first place. The great advantages in this method are its speed, the lack of sophisticated equipment involved and the fact that none of the chemicals involved are particularly unpleasant if handled sensibly. These latter two considerations may be important where pollen extraction is to be attempted in a school environment with unskilled and inexperienced operators.

The main disadvantage with this simple technique is that much extraneous matter still survives in the final suspension which effectively 'dilutes' the pollen on the slide and may obscure it, making identification difficult. In most critical palynological work therefore it is felt that it is worthwhile and, indeed, economical upon time and effort, to process the material further in order to produce a suspension which is easier to count.

In organic peats a major constituent is cellulose, which is present in the cell walls of all vegetable matter within the peat. Fragments of cell debris from leaves and rootlets within the peat form the main source of trouble in samples prepared by KOH digestion alone. This cellulose can be removed by acetolysis.

2. *Acetolysis*

Cellulose is a polysaccharide and can be removed most effectively by acid hydrolysis. The technique described here is basically that of Erdtman (1960). For a peat lacking any siliceous content, the treatment should follow on immediately after KOH digestion and washing (i.e. between steps (f) and (g) of section 1). If hydrofluoric acid treatment is necessary (see section 4) then acetolysis should be used after the hydrofluoric acid processing.

The steps involved in acetolysis are as follows:

(a) Suspend the washed pellet in glacial acetic acid. This dehydrates the organic material. Centrifuge and decant; discard the supernatant.

(b) Add about 6 ml of acetolysis mixture and suspend the sediment. Acetolysis mixture consists of glacial acetic anhydride (or some prefer acetic acid) mixed with concentrated sulphuric acid in a ratio of 9:1. The mixture should be made up freshly each day and, obviously, great care should be taken in the handling of these corrosive materials especially if acetic anhydride is used. Although the mixing of the two liquids does not generate heat to the same extent as that of concentrated sulphuric acid with water, some heat does result and especial care must be taken to ensure that any glassware used is clean and dry. If the mixture is to be discarded it should be poured carefully *into* water. Water should never be dropped into acetolysis mixture. All operations should take place in a fume cupboard.

After adding the acetolysis mixture to the pellet in the polythene (or glass) centrifuge tube, the tube should be placed in a boiling water bath for one minute. Longer time periods may be beneficial, but certain spores show signs of damage after prolonged treatment (i.e. more than three minutes). *Sphagnum* spores are particularly susceptible.

(c) Centrifuge and decant carefully into running water.

(d) Resuspend in glacial acetic acid and centrifuge.

(e) Resuspend in water and centrifuge.

(f) Safranine can again be used for staining, but

some prefer to leave the material unstained. At this stage the pollen exine is coloured pale yellow/brown. Staining, however, does facilitate photography if this is anticipated.

If the material is to be stained, it is advisable to add a few drops of KOH with the final washing since safranine takes effect better in an alkaline rather than an acid medium. Stain and mount as in steps (g) to (j) of section 1.

KOH digestion plus acetolysis is adequate for pollen concentration in the vast majority of organic peats and no further treatment should be necessary. If inorganic particles are present, however, as in most lake sediments (apart from gel muds), rheotrophic peats and soils, additional processing will be necessary, the nature of which depends upon the chemistry of the inorganic particles.

3. Hydrochloric acid treatment

The presence of free calcium carbonate in a sediment can be detected by adding a little dilute HCl, when effervescence results. If this occurs, an excess of 10% HCl should be added to the sediment prior to any other treatment. This should be carried out without heating, since some exine corrosion can result if hot HCl is used.

The suspension may then be centrifuged or neutralized with KOH in preparation for alkaline digestion.

Silica particles in the form of fine and coarse sands and silts require other forms of treatment. Two basic types of process have been developed, in one of which the silica is dissolved and in the other the organic fraction is separated from the siliceous fraction by differential flotation.

4. Hydrofluoric acid treatment

Silica is soluble in hydrofluoric acid, whereas pollen exines and spores are undamaged by it. When necessary, the process can be inserted after KOH digestion, but prior to acetolysis (after step (f) in section 1).

(a) Add about 6 ml (there is no need to measure this accurately) of concentrated (30–40%) hydrofluoric acid (HF) to the pellet in a polythene tube (HF dissolves glass) and suspend with a polythene stirring rod. Place in a boiling water bath for 10 minutes or until no further 'grittiness' can be detected with the stirring rod. Prolonged boiling for an hour or more does not appear to damage the pollen grains. Where much siliceous material is present, the process may have to be repeated with fresh HF. Alternatively, the suspension can be transferred to a nickel crucible and boiled on a sand tray for about three minutes. This should remove all silica.

Emphasis must be placed upon the dangers inherent in the handling of hydrofluoric acid, particularly for the sake of those not familiar with the compound. HF is extremely corrosive on the skin, producing severe ulceration which may take many months to heal. Contact with the skin should therefore be avoided most scrupulously. Rubber or polythene gloves should always be worn when handling the material and these should be checked frequently for punctures. It is sensible to wear an eye-shield in case of splashes. All operations should take place in a Perspex-fronted fume cupboard (glass-fronted ones soon become 'frosted'). Should contamination of the skin occur, wash in running water for 15 minutes and apply a paste consisting of 1 part magnesium oxide to 1·5 parts glycerine. Then seek medical advice. It is sensible to have the paste available wherever HF is being used.

(b) Centrifuge while hot and collect waste supernatant in a polythene container. (Glass plumbing is a common and vulnerable feature of many modern laboratories.) Waste HF can be neutralized with NaOH.

(c) Resuspend the pellet in 10% HCl; warm, but do not boil. This removes silicofluorides which may have been produced during the HF treatment.

(d) Wash twice with distilled water and either stain as in section 1 (g), or continue with acetolysis treatment.

5. *Bromoform flotation*

The principle of differential flotation was first used by Knox (1942) for the recovery of microfossils. The method has been described by Frey (1955) and the following is a summary of his account. The process should be carried out after KOH digestion and washing (i.e. after step 1 (f, see p. 23).

(a) Resuspend the pellet in acetone in order to dehydrate the material, then centrifuge and decant.
(b) Add 5 ml of a mixture of bromoform and acetone with a specific gravity of 2·3. The latter must be checked with a hydrometer and the density of the mixture adjusted by adding acetone (less dense) or bromoform (more dense) as necessary. The contents of the centrifuge tube should be mixed thoroughly and then centrifuged.
(c) The supernatant now contains all particles with a low density, including all pollen grains, spores, sponge spicules, diatom frustules, chitinous remains of arthropods, etc. This is decanted off and the pellet should be treated with a further 5 ml of bromoform and acetone. If necessary, three extractions can be made to ensure the removal of all microfossils.
(d) Microfossils can be recovered from the bromoform/acetone mixture by reducing its specific gravity. This is achieved by the addition of acetone; roughly twice the mixture volume should be added. On centrifuging the microfossils will remain in the pellet and the supernatant can be discarded into a waste bottle.
(e) Wash the sediment in acetone and centrifuge once more.

The material may now be stained and mounted as described in section 1 (g), or it may be acetolysed prior to mounting.

One very obvious advantage of this method when compared with hydrofluoric acid treatment is that siliceous microfossils, such as diatom frustules, can also be recovered, whereas they are dissolved by HF. Also, the avoidance of HF means that specialized facilities such as Perspex fume cupboards are not necessary. However, bromoform can hardly be regarded as a pleasant liquid to handle and a good fume cupboard is still essential for its use.

Practical difficulties are sometimes encountered with this method, particularly that of obtaining a free suspension of fine organic particles in the bromoform/acetone mixture and the avoidance of clumping. Such clumping is especially to be avoided if dense mineral particles are involved, since it can lead to the loss of microfossils.

6. *Oxidation*

In the vast majority of situations, the techniques described above will prove adequate for the production of preparations from suitable sample materials which have a sufficient concentration of pollen for statistically valid counting and in which the pollen should be in a condition suitable for accurate identification. The only other technique used at all frequently is that of oxidation, but this should not prove necessary except in the most difficult sediments. When used it should follow acetolysis. Erdtman and Erdtman (1933) have described the use of chloric oxides as oxidizing agents for the process. Their technique is as follows:

(a) After acetolysis, suspend the pellet in glacial acetic acid (section 2 d above). Add 5–6 drops of sodium chlorate solution and then (carefully) 1 ml concentrated HCl. The reaction is violent and rapid, resulting in the oxidation and bleaching of material in the tube. More than a few seconds' treatment is liable to result in the destruction of pollen grains.
(b) Centrifuge, decant and wash several times.
(c) Stain and mount as in section 1 (g).

The use of various other oxidizing agents, such as nitric acid, hydrogen peroxide, etc. has been described by Gray (1965). Oxidation is particularly effective for the removal of lignin and it also serves to clear microfossils which have become darkened.

Summary

To summarize, pollen preparation techniques should be carried out in the following order:

A. HCl treatment (section 3). Removes free calcium carbonate.
B. KOH digestion (section 1 (a) to (f)). Deflocculates and removes humic colloids.
C. Either HF treatment (section 4) or bromoform flotation (section 5). Removes siliceous material (former) or separates microfossils from bulky mineral material (latter).
D. Acetolysis (section 2). Removes cellulose.
E. Oxidation (section 6). Removes lignin and clears.
F. Staining and mounting (section 1 (g) to (j)).

Certain of these steps (e.g. C and E) may not be necessary, depending upon the original material.

Mounting medium

It has been suggested here that glycerol jelly should be used as a mounting medium for pollen preparations. There is one major objection to its use and that is the fact that grains tend to swell (Faegri and Deuse, 1960). This makes it difficult to employ size criteria for identification. An alternative medium is silicone oil (Andersen, 1960), which has a low refractive index (see Berglund *et al.*, 1960), a high viscosity (2000 plus centistokes), and does not appear to cause swelling (glycerol jelly mounted grains are usually 1·25 times larger). The use of silicone oil, however, does necessitate dehydration prior to mounting. This is carried out as follows:

(a) Wash with 96% alcohol, centrifuge and decant.
(b) Wash with 99% alcohol, centrifuge and decant.

(c) Wash with benzene, centrifuge and decant.
(d) Add 1 ml of benzene and transfer to an open dish. Add silicone oil, mix and allow to stand for 24 hr, during which time the volatile benzene will evaporate. This is best performed in a fume cupboard since benzene fumes are harmful to health.
(e) Finally dilute with silicone oil to the required microfossil concentration and mount. Sealing of slides is to be recommended and they should be stored horizontally.

A general discussion of mounting media and techniques is given in Andersen (1965). Davis (1966) has suggested the use of tertiary butyl alcohol as an alternative to benzene.

Type slides

Although keys and photographs are a valuable aid to the identification of pollen grains, confirmation of identity requires comparison with modern pollen. It is essential for anyone involved in pollen analysis to have access to, or to build up for himself, a type collection of modern pollen grains.

Only two points need to be stressed with respect to type pollen collections. The first is that accurate taxonomy is necessary when preparing samples from fresh or herbarium material. The second point is that type material should be prepared by the same methods and mounted in the same material as that regularly used for subfossil samples. This ensures that the comparison of fossil with type material is a valid one. Potassium hydroxide treatment, followed by acetolysis, staining in safranine and mounting in glycerol jelly produces good results with fresh material, but slides should be sealed before storage. If various stains and mountants are used in preparing fossil material then it is advisable to prepare several typeslides of each species using the different stains and media.

Microscopy

It is possible to identify a large number of pollen types even with the use of unsophisticated equipment. Precise work, however, necessitates high grade microscopy. One essential for quantitative counting of microfossils is a mechanical stage, since this permits regular scanning and traversing of slides, thus avoiding crossing the same area more than once. Also highly desirable is a binocular head, since monocular attachments can result in eyestrain after long periods of pollen counting.

Scanning is best carried out at a magnification of $\times 400$ to $\times 600$, but precise identification of difficult taxa may require magnifications of over $\times 1000$ and the use of oil immersion lenses. Although phase-contrast equipment is not essential for crude analyses, it is highly desirable for precision work. In the key (Chapter 5) reference has occasionally been made to characters visible only with the aid of phase contrast microscopy where this technique offers the best opportunity of making certain distinctions.

Absolute pollen extraction techniques

All the techniques of pollen extraction described so far in this chapter supply material which can be used for the determination of only relative frequency of different pollen types. They do not provide a means of estimating the *density* of that pollen type within the sediment. In certain circumstances a knowledge of densities can be extremely informative, for example if the rate of sedimentation is known it permits the calculation of the absolute influx rate of the pollen type in terms of number of grains per unit area of sediment surface per unit time. The interpretation of such information and its value for pollen analysis is discussed in Chapter 6.

As far as extraction of pollen is concerned, the main problem is the development of a technique which will allow the determination of the absolute density of pollen grains in a sample of known volume which occupies a known vertical depth of sediment. This latter requirement is necessary for the calculation of how long the sample took to accumulate once the sedimentation rate is known.

Several techniques have been used to determine pollen densities. Very early in the history of pollen analysis von Post (1916) used an absolute method by treating and counting a known volume of sediment. At that stage, however, no methods were available for the determination of sedimentation rates and major variations in absolute pollen densities with depth could often be accounted for by changes in sediment deposition rate. For this reason von Post's absolute methods were abandoned in favour of conventional relative counts in which density estimates are not required.

With the availability of radiocarbon dating as a means for estimating sedimentation rates, interest has been revived in density counts. Techniques have been of three basic types:

1. Volumetric methods (e.g. Davis and Deevey, 1964; Davis, 1965; 1966).
2. Weighing methods (e.g. Jørgensen, 1967).
3. Exotic marker grain methods (e.g. Benninghof, 1962; Bonny, 1972).

Techniques 1 and 3 necessitate the accurate determination of sample volume prior to treatment. This is a difficult process and several techniques have been attempted in order to achieve repeatable results. Davis (1969) used a spatula of exactly 1 cm³ capacity and filled this with the sample. This method suffers from the disadvantage that the core of sediment is easily distorted or compressed, thus changing its density. A boring device which produces samples which are apparently free from differential compaction has been designed and described by Engstrom and Maher (1972). Bonny (1972) determined subsample volumes by displacement in water.

In weighing methods, the sample weight rather than its volume is required.

Volumetric methods

The sample of known volume is subjected to the pollen concentration processes described earlier in the chapter. The sample is then dehydrated in benzene (Davis, 1965) or tertiary butyl alcohol

(Davis, 1966) and is suspended thoroughly in a known volume of the liquid. Aliquots, also of known volume, can then be removed by means of a pipette and are mounted in silicone oil. Quantities of material involved in these preparations are adjusted to produce a convenient concentration of pollen on the final slides so that all pollen grains can be counted.

The use of tertiary butyl alcohol rather than benzene in this preparation is advantageous both because of its lower volatility and also because it renders pollen grains less liable to stick to one another or to the glass (for example, of the pipette) during transference to slides.

Weighing methods

In these methods (e.g. Jørgensen, 1967) a known weight of sample is subjected to pollen concentration. Finally the preparation is thoroughly suspended in a known weight of glycerol. A weighed subsample of this suspension is then mounted and all pollen present on the slide is counted. It is then possible to calculate the concentration of pollen in terms of the weight of the original sediment.

Both of these methods, volumetric and weighing, require the counting of all grains in the final preparation and this means that the final concentration must be judged carefully to produce a reasonable number of grains for convenient counting. This problem is overcome in the third type of absolute pollen counting technique.

Exotic marker grain methods

This was first employed by Benninghoff (1962) and has subsequently been modified by Matthews (1969); Pennington and Bonny (1970) and Bonny (1972).

A requirement for the technique is a suspension of known concentration of a pollen type which is of an appropriate size, which does not agglomerate and hence which disperses easily in the suspension medium (e.g. glycerol), which is unlikely to occur naturally in the sediment under investigation (Bonny used *Ailanthus glandulosa*),

and which is easily recognizable. The homogeneity of the suspension must be checked very thoroughly and the concentration of pollen determined accurately.

A sediment sample of known volume is subjected to pollen concentration procedures. The final pellet is weighed and then a weighed quantity of exotic suspension is added, the precise quantity of which is determined by the consideration that the final concentration of exotic pollen as a proportion of total pollen should be 20–40%. From the weight of the suspension added, the number of exotic grains added can be calculated. Subsamples are mounted on slides.

Counting is carried out on replicated subsamples and the concentration of each pollen type relative to the exotic concentration is calculated from the sample counts. It is thus not necessary to count all pollen grains on a slide. From these data one can calculate back to the original concentration of pollen types in the sediment.

Maher (1972) has modified this method by adding his suspension of exotic pollen before the pollen concentration procedures. Also he used a volumetric rather than weight basis for the addition of a known quantity of exotic pollen. Matthews (1969) also added his suspension prior to pollen concentration. He was concerned about the flotation properties of certain exotic types such as *Liquidamber styraciflua* in 10% KOH. It is possible that acetolysis of the exotic prior to its use would overcome such a problem. He found that *Nyssa sylvatica* and *Ailanthus elegantissima* if defatted did not behave in this way. Their proportions remained the same throughout processing. Cushing (see Craig, 1972) adds pollen-sized polystyrene spheres as markers to his sediments and a similar method, using plastic spheres has been used by Craig (1972).

These methods and modifications overcome many of the problems inherent in volumetric and weighing methods and are the most widely used means of determining the absolute density of pollen grains in sediments. A comparison of the relative merits of these techniques has been made by Peck (1974).

Chapter 4
Pollen Grains and Spores

In angiosperms and gymnosperms the wall of the pollen grain has the important function of protecting the male gametophyte in its journey between anther and stigma. In lower plants such as pteridophytes and bryophytes the spore has the function of dispersal of the plant to suitable moist habitats where the gametophyte can grow. Pollen grain and spore walls are therefore very resistant to water loss and are often elaborately sculptured. It might be that a resistant wall is needed primarily to avoid desiccation during the aerial journey. This explanation does not, however, account for the elaborate and distinctive sculpture that is found, nor for the enormous range of sculpturing types so useful to the pollen analyst. The adaptive significance of these sculpturing types is still largely unknown, but many ideas have been put forward during recent studies of the ultrastructure of pollen grain and spore walls (Heslop Harrison, 1971b).

Structure and composition

In the living pollen grain of an angiosperm, the wall is made up of two layers; the outer layer is called the *exine* and is composed of a very unusual substance, *sporopollenin* (Zetsche, 1932). The inner layer, or *intine*, is made up of cellulose and is very similar in construction to an ordinary plant cell wall. In fossilization only the very resistant sporopollenin-containing exine remains, and it is this that carries all the morphological characters necessary for pollen recognition. For a long time so little was known about sporopollenin that it was defined by its resistance to various methods of chemical attack, especially to the process known as acetolysis. It can be affected by oxidants, but it requires very strong oxidation (e.g. with sulphuric acid mixed with hydrogen

peroxide, or treatment with 40% chromic acid or ozone) to be completely broken down. It is affected by only two bases, fused potassium hydroxide and 2-aminoethanol. In 1968 Brooks and Shaw suggested that sporopollenin was a complex polymer of carotenoids and carotenoid esters with oxygen. They found that these two

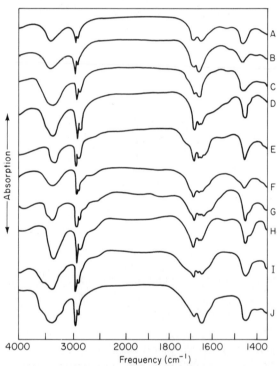

Fig. 4.1 The infra-red spectra of some sporopollenins (after Brooks and Shaw 1971). A. *Chara corallina* algal spore. B. *Pediastrum duplex* algal spore. C. *Mucor mucedo* (±) fungal spore. D. *Lycopodium clavatum* spore exine. E. *Lilium henryi* pollen exine. F. Oxidative polymer from *L. henryi* carotenoids and carotenoid esters. G. Oxidative polymer of β-carotene. H. *Selaginella kraussiana*, a modern megaspore. I. *Valvisporites auritus* a fossil megaspore (250×10^6 yr old). J. *Tasmanites punctatus* fossil spore exine (350×10^6 yr old).

types of compound were the dominant lipid material formed in the early stages of anther growth in the lily (*Lilium henryi*). Carotenoids extracted from lily anthers were artifically polymerised, and the infra-red spectrum of the substance produced was remarkably similar to the spectra of sporopollenins from a wide variety of sources (Fig. 4.1). The work of Brooks and Shaw also shows that sporopollenin is much more widespread than was originally thought. The spectra show that it is found in the spores of such widely separated plant groups as algae, fungi, pteridophytes and angiosperms. Perhaps here we have parallel evolution of this propagule-protecting substance (Heslop Harrison, 1971b). Studies of some of the oldest sedimentary rocks in the world, the Precambrian rocks of the Overwacht (3.7×10^9 yr old) and the Fig Tree cherts of South Africa (3.2×10^9 yr old), have shown the presence of amorphous insoluble organic material which appears similar to present-day sporopollenin (Brooks and Shaw, 1968; 1971). The presence of this unique plant polymer in such very early sediments has important implications for investigations into the origin of life. Brooks and Shaw have also found small amounts of a sporopollenin-like substance in the organic matter of the Orgueil and Murray meteorites. They believe that this is powerful evidence for the existence of extra-terrestrial life.

Although sporopollenin is so widespread, it is only in the higher plants (angiosperms and gymnosperms) that it is built into the complex wall structures evident in the photographs in this book. Much work has been carried out on angiosperm pollen grains and their structure and genesis is known in some detail. The terminology used differs among various authors, hence some definitions are necessary. The exine of pollen grains is divided into an outer sculptured *sexine*, and an inner unsculptured *nexine* (Erdtman, 1966). The sexine commonly takes the form of a set of radially directed rods, supporting a roof. This roof may be complete, partially dissolved or completely absent. Following the suggestions of Reitsma (1970) we shall call the roof a *tectum*,

a rod which supports a tectum or any part of it, a *columella*, and a rod which is not supporting anything, a *baculum* (see Fig. 4.2 system B). The term 'baculum' was first coined by Erdtman and was used to mean any rod in the sexine, whether it was supporting something or not: it is used in this sense in many of the papers referred to later on.

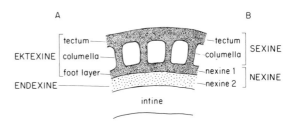

Fig. 4.2 Terminology of the pollen grain exine. A. As defined by Faegri (1956), Faegri and Iversen (1964). B. As defined by Reitsma (1970). Faegri has stressed the difference in the staining characteristics of the ektexine and the endexine. Reitsma's division (following that of Erdtman, 1966) is a formal morphological one, dividing the exine into convenient descriptive units.

System A in the diagram differs from B in that it derives from the various staining characteristics of the layers of the exine, and not the sculpturing differences of the layers. If the wall is stained with basic fuchsin the 'ektexine', which is actually the sexine plus nexine 1, becomes heavily stained before the 'endexine' or nexine 2 (Faegri, 1956). Recently Southworth (1974) has shown that the ektexine of some pollen grains is soluble in hot 2-aminoethanol and several related substances while the endexine is left unaffected. The staining and solubility differences indicate that the sporopollenin of these two layers may differ chemically. In addition, Afzelius (1956) in an early electron-microscopical study found a difference in the way sporopollenin had been deposited in the layers of the exine. The endexine had a laminated appearance while ektexine was amorphous–granular. Some recent studies show that this may indicate a difference in the consolidation of the sporopollenin in the two layers.

The sporopollenin of ektexine appears laminated in a radial direction when it is first laid down, but this disappears later as it consolidates. The endexine sporopollenin is deposited in tangential layers and retains this lamination when the grain is mature; the endexine is thus less consolidated than the ektexine.

Despite the fact that chemical and developmental studies of the pollen grain wall indicate that a division into ektexine and endexine is the most basic one, distinguishing these layers with a light microscope presents problems. A foot layer (nexine 1) is not always present and even when it is found, it is often very difficult to distinguish from the endexine (nexine 2). With the light microscope it is easy to distinguish between sculptured sexine and unsculptured nexine, thus these are better terms from the point of view of the pollen analyst, whose main concern is describing the characteristics of a pollen grain exine in order to identify the grain or to compare its characteristics with those of other grains. For this reason we shall adhere to the sexine-nexine division for the rest of this book.

Although gymnosperms appear to have a wall stratification similar to that of angiosperms (layers have been tentatively assigned to sexine and nexine), spores of pteridophytes and bryophytes do not resemble angiosperms in their wall structure. The walls of their spores often appear to be laminated throughout their thickness and bear no layer containing columellae. In pteriodophytes a division on a different basis has been attempted, for there is often a loose outer layer surrounding an inner layer. The outer is called the *exosporium* (or *perine*) and the inner layer the *endosporium*.

The development of the pollen grain wall

The method whereby the pollen cell can produce the complex wall which surrounds it is a subject that has fascinated people for many years. Just after the turn of the century, pollen grain development attracted the attention of the cytologist Rudolf Beer and, although a few other

studies had been made previously, his was the most significant of the early works. In 1911 he presented an account of the development of the pollen grain wall of the morning glory (*Ipomoea purpurea*), a species possessing large pollen grains with elaborately sculptured walls. In the anther, each *pollen mother cell* undergoes meiotic divisions to produce a tetrad of four haploid *microspores*. Beer saw that the main features of the sculpturing pattern appeared on each microspore while it was still enclosed in the callose wall of the pollen mother cell, so the patterning was not produced as a result of contact with any of the parent plant tissue in the anther. Beer also observed fibrils running from the nuclear membrane towards the cell wall during this pattern-determining period. They disappeared after the release of the spores from the tetrad, but the pattern remained. As far as he could see this meant that it was the haploid microspore nucleus which determined the template of the mature pattern. This was as far as the story could be taken with the light microscope; the details of the process had to await the invention of the transmission and the scanning electron microscopes. Since the diversity of pollen types is wide, one would expect there to be differences in the wall construction process, and this has been found to be the case. Excellent reviews of the recent studies on the pollen wall development of various species are given by Echlin (1968), Godwin (1968) and Heslop-Harrison (1971a). These form the basis of the generalized account given here.

The process of pattern formation begins early in the tetrad period, as soon as each cell is independent and completely isolated from the other cells by a thick callose wall. A cellulose layer is then formed between the plasmalemma of the microspore and the investing callose wall. This cellulose layer forms a continuous sheath to the grain, except over those areas which are destined to be apertures (pores or furrows) in the mature grain. In these areas a plate of endoplasmic reticulum is closely apposed to the plasmalemma, which in turn lies directly against the callose tetrad wall. Thus the endoplasmic reticulum acts as a stencil, allowing cellulose to be deposited everywhere except those areas it has 'blocked'. The initial positioning of these endoplasmic reticulum plates is often correlated with the position of the microspores within the tetrad (Wodehouse, 1935). The common tetrahedral disposition of the spores (see Fig. 4.5) means that each spore has three faces of contact with its sibs. It is on these three faces that apertures are often formed, e.g. the pores in three-pored (triporate) grains. If the 'contact geometry' is destroyed, aperture formation is disrupted. In the lily, Heslop-Harrison has shown that when colchicine is used to block the two meiotic divisions of the mother cell, the whole mother cell then behaves as a single spherical spore in that either apertures are not formed or one irregular, randomly placed aperture is formed.

The cellulose layer in the non-apertural areas is soon seen to be traversed by radially directed rods named *probacula* by Heslop-Harrison (in our terminology they would be called procolumellae). These probacula at first appear to be lamellated in a radial direction and are composed of lipoprotein. Thus before any sporopollenin has come on the scene the three major characters of the future exine have appeared, the number and position of the apertures and the distribution of the bacula (columellae). From some investigations, e.g. Skvarla and Larson (1966), the sites of the probacula have been thought to be determined by some cytoplasmic membranes in a manner reminiscent of Rudolf Beer's 'fibrils'. In other investigations these structures have not been found. As to the question of whether the haploid nucleus controls the patterning of the probacula, the basic 'programme' for exine pattern has been shown to be present in the cytoplasm even before cleavage of the mother cell (Heslop-Harrison, 1971b). Thus any nuclear genes coding for the pattern must be transcribed in the diploid mother cell, not in the haploid microspore. This has been shown by disrupting normal cleavage and producing fragments of cytoplasm without a nucleus. These fragments go on to produce a normally stratified exine with the correct pattern of columellae.

After the appearance of the probacula, connections develop between their feet to form the future nexine 1 (foot layer), and between their heads to form the future tectum. The probacula become much more electron dense and their lamellated structure disappears. This is the time when the probacula become resistant to acetolysis so it is assumed that sporopollenin deposition has occurred. This is called the 'primexine' stage. Fig. 4.3, A shows approx-

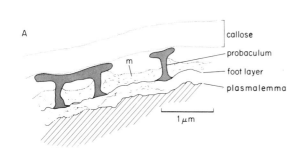
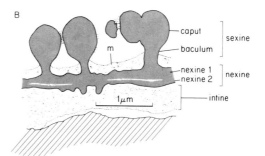

Fig. 4.3 Drawings from electron micrographs of *Lilium longiflorum* pollen grain exines (after Heslop-Harrison, 1968).

A. Late primexine. Here the probacula are visible as electron-opaque rods with linking muri above them and matrix material (m) between them. The foot layer (nexine 1) is forming from the bases of adjacent probacula. The gap between this and the plasmalemma may be an artifact due to fixation. Note callose wall which envelopes microspore.

B. Mature exine. Here each baculum has a swollen head and the foot layer has fully developed, becoming divisible into nexine 1 and nexine 2. The indefinite layer below nexine 2 is the intine, which is composed of cellulose only and is the last of the walls to appear.

imately this stage in the lily pollen grain. Shortly after this the thick callose wall breaks down, and the pollen grain is released from the tetrad. On release the grain undergoes a rapid size increase, causing the primexine to stretch and become thinner. Consequently the cellulose layer (in which the probacula are embedded) is shredded and dispersed, leaving only a remnant lying on the foot layer between the probacula. The thinning of the primexine is not as noticeable as it might be because sporopollenin accretion is occuring all the while, thickening the probacula, tectum and foot layer (see Fig. 4.3, B).

The formation of nexine 2 (endexine) and the cellulose intine begins slightly before the dissolution of the callose wall and the break-up of the tetrad. Nexine 2 is built up by the apposition of tangentially orientated lamellae, produced at the plasma membrane and upon which sporopollenin has been deposited. These lamellae resemble those seen in the early probacula. In some species the lamellated appearance of this layer persists in the mature grain. Rowley (1963) showed that this laminated material forms a large part of the thickenings around the pore of grass pollen grains. Thus nexine 2 can contribute to the architecture of the grain by variations in its thickness.

When considering the part that parent plant tissue plays in the formation of the exine it is interesting to look at the origins of the sporopollenin in the exine layers. The sporopollenin of primexine and nexine 2 are thought to come from within the haploid spore. At this time the callose wall cuts off all exchange of materials with the anther cavity and the other spores. However, after dissolution of the callose wall the spore is set free in the anther space, which often means close contact with the tissue known as the tapetum. Within the tapetal cells are small granules or droplets of what appears to be lipid material. These have been called *orbicules* or *Ubisch bodies*. Brooks and Shaw (1971) believe that this lipid material consists of carotenoids and carotenoid esters, the precursors of sporopollenin. Many fine-structural studies, especially

those of Rowley (1963) and Echlin and Godwin (1968) show that these orbicules are released into the anther fluid as the walls of the tapetal cells break down, and on release they seem to gain a sporopollenin skin. The role of these orbicules is not yet clear; one view is that they actually come into contact with the spore exine and transfer the sporopollenin skin to the structures of the developing exine. Banerjee and Barghoorn (1971) have found that the orbicules appear to form the outermost spinules on the tectum of some grasses. Another view is that they have no role in sporopollenin transfer and just happen to become coated with sporopollenin because, like the pollen grains, they are free in the anther fluid and sporopollenin deposition is going on over all available surfaces.

One thing that is certain is that many other substances are coated on to the mature pollen grain from the disintegrating tapetum. These substances have been given the names tryphine or 'pollenkitt' and they include the characteristic pigments and sticky or odorous materials that are important in the pollination process. The tapetum imparts yet another class of compounds to the outer part of the exine (and to the spaces between the columellae) of some pollen grains, namely the 'recognition' proteins which ensure that the pollen grain germinates only on a compatible stigma. If the stigma is incompatible the recognition proteins instigate a reaction which stops pollen tube growth.

Application of morphology to the identification of fossil pollen grains and spores

In the study of pollen grains and spores, the complexity of their structure and patterning has necessitated a formidable terminology. In addition, structures have been given different names by different investigators. We have already come up against this sort of problem in the naming of the layers of the exine (see page 31). Unfortunately there is now a situation where the two major works for pollen grain identification (Faegri and Iversen, 1975; Erdtman *et al.*, 1961;

1963) use different terminologies. We shall here adopt the suggestions of Reitsma (1970), who has attempted to unify these terminologies. All the terms we think it necessary to use are in the glossary on page 74.

Apertures

For the identification of any fossil grain or spore the first features to note are those of the apertures. An *aperture* is any thin or missing part of the exine which is independent of the pattern of the exine, e.g. in *Buxus* (Plate 14) the apertures could not be confused with the holes of the reticulate pattern because the apertures are much bigger and cut across the pattern. There are two sorts of aperture named *pori* (pores) and *colpi* (furrows). Colpi are thought to be more primitive than pori and are distinguished from the latter by being long and boat shaped with pointed ends. Pori are usually isodiametric holes, but can be slightly elongated with rounded ends. Grains with pori are called *porate*; with colpi, *colpate*; and with both colpus and porus combined in the same aperture, *colporate*. In the living grain apertures are not actually open, but are covered by a thin and delicate layer of exine material. In contrast, the intine found under apertures is usually thicker than that elsewhere on the grain. An aperture is usually thought of as being the site of emergence of the pollen tube at germination on a compatible stigma. As only one aperture is used for pollen tube exit, the large number of apertures found on some grains (e.g. up to fifty in the Malvaceae) leads to the idea that they must have some additional function. The thick intine region under the apertures has been found to have a store of easily leached proteins which may function in the recognition reaction between pollen grain and stigma (Heslop-Harrison, 1971b). Apertures are thus also exit sites for these proteins. Wodehouse (1935) was the first to suggest that apertures were regulatory devices, controlling the movement of water into and out of the grain. Dehydration will cause an infolding of the intine at a colpus, and its margins will come close together. This reduces the area through which water can be lost (a negative feedback effect). Imbibition of water causes a gaping of the colpus margins and the covering membrane is stretched, thus increasing the area through which water can enter (a positive feedback effect). Pori operate in the same way, but it is less noticeable.

An aperture (colpus or porus) is demarcated by a line most often caused by changes in the thickness of the sexine or nexine or of both at once. Thus it can be said to be situated in the sexine or in the nexine or in both. In the detailed analysis of pollen grain structure an aperture which is a feature of the sexine is called an *ectoaperture* (*ectocolpus*, *ectoporus*) and an aperture which is a feature of the nexine is called an *endoaperture* (*endocolpus*, *endoporus*). In some cases ecto- and endoapertures are of the same type (i.e. pori or colpi) and occur in the same place, in other cases they may be of different types occuring in slightly different positions. For example in *Centaurea cyanus* [Fig. 4.6 (E i) and (E ii)] there are three ectocolpi running meridionally and one continuous endocolpus running equatorially and forming a complete girdle to the grain. Other examples are *Polygonum aviculare* where there are three meridional ectocolpi each crossed by an equatorial endocolpus and *Oxyria* or *Fagus* where there are three meridional ectocolpi each underlain by an equatorial endoporus.

Pollen grains and spores can be divided into groups on the basis of the number, position and character of their apertures. The classification is basically simple and consistent. The number of apertures is indicated by attaching the prefixes *mono-*, *di-*, *tri-*, *tetra-*, *penta-* and *hexa-* before the terms *colpate, porate* and *colporate*. More than six apertures is indicated by the use of the prefix *poly-*. In most cases the pori and/or colpi are arranged equidistantly around the equator of the grain. This situation is indicated by the prefix *zono-*. If the apertures are scattered all over the surface of the grain the prefix *panto-* is used.

For example:

Polyzonoporate—with more than six pores, these pores being situated in equatorial zone.

Polypantoporate—with more than six pores, these pores being scattered all over the surface of the grain.

Pentazonoporate—with five pores arranged in an equatorial zone.

Pentapantoporate—with five pores scattered all over the surface of the grain.

Fig. 4.4 shows the whole range of pollen types possible. There are some types which do not fit neatly into the system outlined above. One of these is the *syncolpate* grain. Here two or more colpi may fuse, usually at the poles of the grains

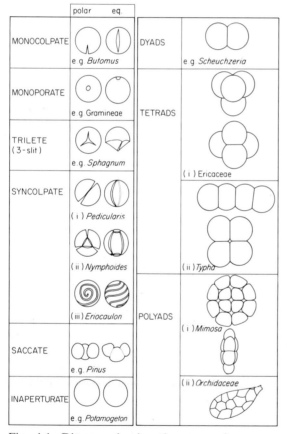

Fig. 4.4 Diagram showing the range of aperture number, position and character. Some of the possible combinations have no example within the British flora. (Continued on facing page).

(e.g. *Nymphoides*, *Pedicularis*, Plates 15 and 16) but occasionally elsewhere. An example of the latter case is given by *Eriocaulon* (Plate 14) where a set of colpi are fused, giving the appearance of a set of spirals surrounding the whole grain. Another odd grain is that of the members of the Compositae subfamily Liguliflorae. Here there is a simple aperture system (tricolporate) but it is obscured by the very unusual sexine pattern. There are large apparent gaps in the sexine which are separated by high spiny ridges (e.g. *Taraxacum*, Plate 2). We have retained the name that Faegri and Iversen gave to this grain—*fenestrate* (with windows). In some genera or families the pollen grains are characteristically present as aggregates, e.g. *tetrads* in the Ericaceae or Typhaceae and *polyads* in the Orchidaceae.

Spores of pteridophytes and bryophytes possess very different wall structure to angiosperms and gymnosperms (see page 32). As there is no sexine–nexine division any aperture that they possess cannot be assigned strictly to colpi and pori, as these are sexine features. Some spores have one long slit-shaped aperture, e.g. Polypodiaceae, while others have a three-branched slit forming a Y shape, e.g. *Sphagnum*. These various slit marks on spores are actually tetrad break-up scars. Despite the fact that they are a relic of the tetrad stage, these slits are in no way related to the aperture system of angiosperms. The slits always occur at the proximal pole of the spore whereas angiosperm pori and colpi occur on the distal pole, or variously over the surface but rarely on the proximal pole. Fig. 4.5 shows a tetrahedral tetrad and the way the parts of a grain or spore are named according to the orientation of that grain or spore in the tetrad. Once separated from the tetrad it is often impossible to tell the orientation of a grain and thus it is very difficult to determine whether an aperture is on the distal or proximal face, and consequently whether it is, for example, a true colpus or a slit in a spore. For this reason all one-slit spores have been included under the heading *monocolpate* in the key. For three-slit spores the problem does not arise because their aperture

Fig. 4.4 *(contd)* Classification of pollen types based upon the number and arrangement of apertures. Examples are shown in polar and equatorial views. Dotted lines indicate a different focal plane. Empty positions denote the lack of a north-west European example.

type is almost unique, they therefore have a separate class on their own (*trilete*).

Those areas on a grain which are not occupied by apertures are given names depending on whether they are adjacent to pori or colpi. The area bordered by two colpi is called the *mesocolpium*, and that bordered by two pori is called the *mesoporium*. If the pori or colpi are in the zono-arrangement, at each pole there is an area where no apertures occur. This polar area is called the *apocolpium* if the zonally arranged apertures are colpi, and the *apoporium* if the zonally arranged apertures are pori (see Fig. 4.5, B).

G). Thickenings of the nexine around an endoaperture [*Centaurea cyanus*, Fig. 4.6, E (i) and E (ii)] or below the edge of an ectoaperture (Fig. 4.6, A) are called *costae*. Thinnings of the nexine are also known (e.g. *Myrica* Fig. 4.6, C). In some grains the two layers of the exine become separated from one another in the vicinity of the apertures. The cavity so formed is commonly found around pori (e.g. *Betula* Fig. 4.6, B) and is called a *vestibulum*. Other grains have the central part of the aperture membrane with a sexine layer as thick as that occurring on the main body of the grain. This thickened centre is

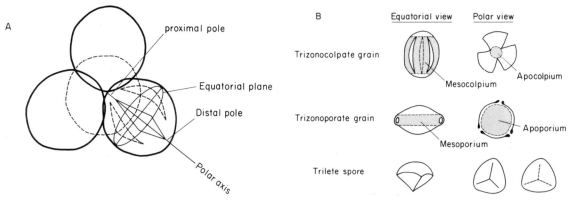

Fig. 4.5　A. Tetrahedral tetrad showing how the parts of a grain are named. The proximal pole of a grain is that which was nearest the centre of the tetrad, the distal pole that which was farthest away from the centre of the tetrad. The polar axis is a line which passes from the centre of the proximal pole to the centre of the distal pole. The tetrad also shows how aperture position (in this case three colpi) can be influenced by the orientation of the grain within the tetrad.
B. Examples of how grains with differing aperture types would appear in polar and equatorial view. The diagrams also show how the apertures partition the surface of the grain into areas which are given names for descriptive purposes.

Having dealt with the number and position of the apertures, the next feature to note is their character. The exine often shows a slightly altered structure in the vicinity of apertures. When this happens, the aperture is said to be *bordered*. Borders can be a feature either of the sexine or of the nexine, and a few examples of them are shown in Fig. 4.6. A sudden thickening or thinning of the sexine around an ectoporus is called an *annulus* (Fig. 4.6, C, D), and around an ectocolpus is called a *margo* (e.g. *Salix*, Fig. 4.6,

called an *operculum* and can be seen on the pori of *Plantago lanceolata* (Fig. 4.6, D), and on the colpi of *Potentilla* type. As the membrane is still very thin around the edges of the aperture, opercula are frequently lost during fossilization (e.g. as in grasses).

Shape

Sometimes the shape of a grain or spore is useful in identification. However it is wise not to lay too

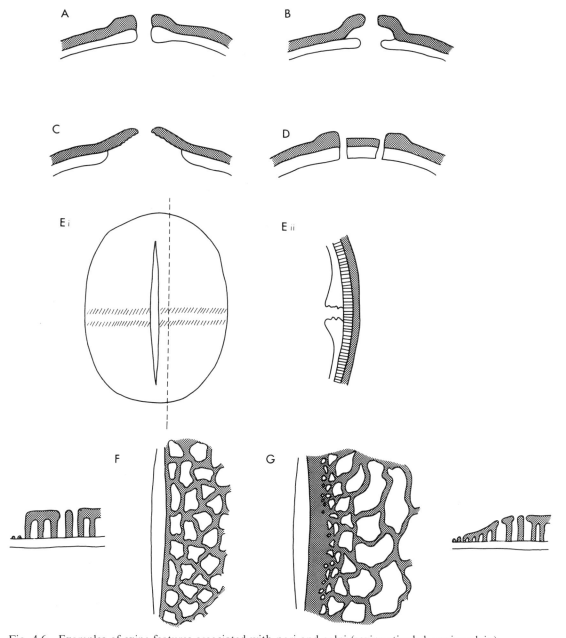

Fig. 4.6 Examples of exine features associated with pori and colpi (sexine stippled, nexine plain).

A. Porus with a *costa* (thickening of the nexine), e.g. Gramineae.

B. Porus where sexine separates from nexine to form a *vestibulum*, e.g. *Betula*.

C. Porus with an *annulus* formed by a slight thickening of the sexine, e.g. *Myrica*. The nexine is here absent in the vicinity of the porus.

D. Porus with an *operculum* (thickening of the middle of the aperture membrane) and an annulus (sexine thickening), e.g. *Plantago lanceolata*.

E(i). Equatorial surface view of *Centaurea cyanus* to show *equatorial endocolpus*.

E(ii). Section of the exine along the dotted line in E(i). This shows the *endocolpus* to be bordered by heavy thickenings (costae) of the nexine.

F. Colpus without a border or *margo*, i.e. the lumina of the reticulum remain the same size right up to the colpus edge. The sexine also remains the same thickness right up to the colpus edge, e.g. *Fraxinus*.

G. Colpus with a *margo*, i.e. the lumina become smaller towards the colpus edge and disappear at the edge, giving a tectate margin. The sexine becomes gradually thinner towards the colpus edge as is shown in the section, e.g. *Salix*.

great an emphasis on the shape because it can vary considerably within one grain type or even within one species. Variation in shape is also caused by the choice of extraction methods, and embedding media. Although pollen grains and spores are three-dimensional objects, this fact is often difficult to observe using the light microscope, where only one plane is in focus at any time. Thus pollen grains and spores are described by the shape of their outline in polar

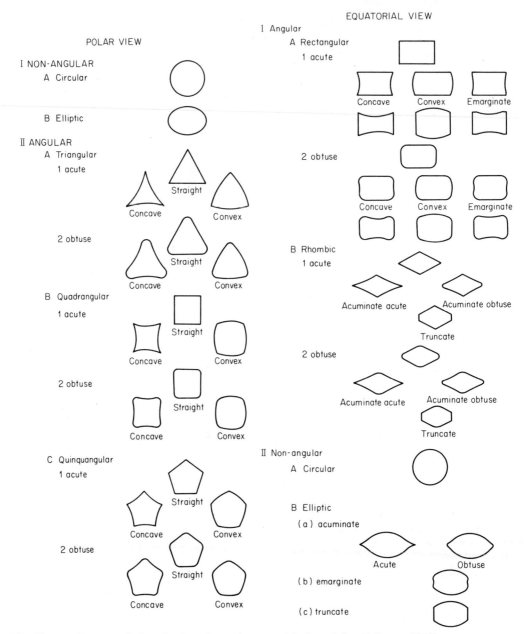

Fig. 4.7 Shapes of symmetrical grains in polar and equatorial views (after Reitsma, 1970).

TABLE 4.1   *Description of sculpturing types*

Type	Description	Example	Plate reference
Psilate	With surface completely smooth	*Aconitum*	20
Perforate	With surface pitted (holes dark at high focus). Pits < 1 μm	*Convolvulus*	20
		Cerastium	13
Foveolate	With surface pitted (holes dark at high focus). Pits > 1 μm	*Fagopyrum*	36
		Tilia	38
Scabrate	With projecting elements isodiametric, no dimensions greater than 1 μm (appear white at high focus)	*Thelycrania*	32
Verrucate	With width of projecting elements as great as height (appear white at high focus)	*Plantago major*	12
Gemmate	With width of projecting elements same as height, but elements constricted at bases (appear white at high focus)	*Nymphaea*	5
Clavate	With height of projecting elements greater than width, bases constricted (appear white at high focus)	*Ilex*	29
Pilate	With height of projecting elements greater than width, apical parts of elements swollen (appear white at high focus)	*Mercurialis*	47
Baculate	With height of elements greater than width, bases not constricted (appear white at high focus)	*Linum*	15
Echinate	With projecting elements pointed	*Malva*	14
		Lonicera	24
		Campanula	10
Rugulate	With projecting elements elongated sideways, length 2 × breadth and irregularly distributed (white at high focus)	*Sedum*	45, 46
		Nymphoides	15
		Prunus	45
Striate	With projecting elements elongated sideways and arranged ± parallel to one another	*Menyanthes*	28
		Acer	27
		Potentilla	45
Reticulate	With projecting elements arranged in a network pattern (net appears white at high focus)	*Cruciferae*	22
		Iris	17
		Potamogeton	4
		Lysimachia	39

and equatorial views. Fig. 4.5, B shows some examples of polar and equatorial views. Fig. 4.7 shows the naming of shape classes that we shall use.

Sculpturing

After separating pollen grains and spores into classes based on the features of their apertures, these classes can be divided by consideration of the fine structure and pattern of the exine. The sexine has been described as being composed of small radially directed rods which sit on the nexine and are called columellae if they support something (e.g. the tectum, a plate or a small knob) and bacula if they do not support anything and are cylindrical in shape. In some cases the rods are obviously free at their heads, but are

non-cylindrical in shape. In such cases they are called *clavae* if they are club-shaped, *echinae* if they are sharply pointed, *pila* if they have swollen heads and *gemmae* if they are short and globular. Sometimes the sexine elements which sit on the nexine do not resemble rods at all. They may appear to be small hemispherical warts (*verrucae*) or tiny flakes (*scabrae*) or other small elements (*granules*). In Table 4.1 there is a more detailed description of all these sexine elements. When the heads of the columellae are joined by a complete tectum, the grain is described as *tectate*. The columellae are usually simple, but in some genera, e.g. *Convolvulus* (Plate 20) and *Geranium* (Plate 9), they branch in a quite complex manner. A grain with free rods, i.e. bacula, clavae, echinae, etc. is described as *intectate*. Tectate and intectate are two extreme conditions and grains

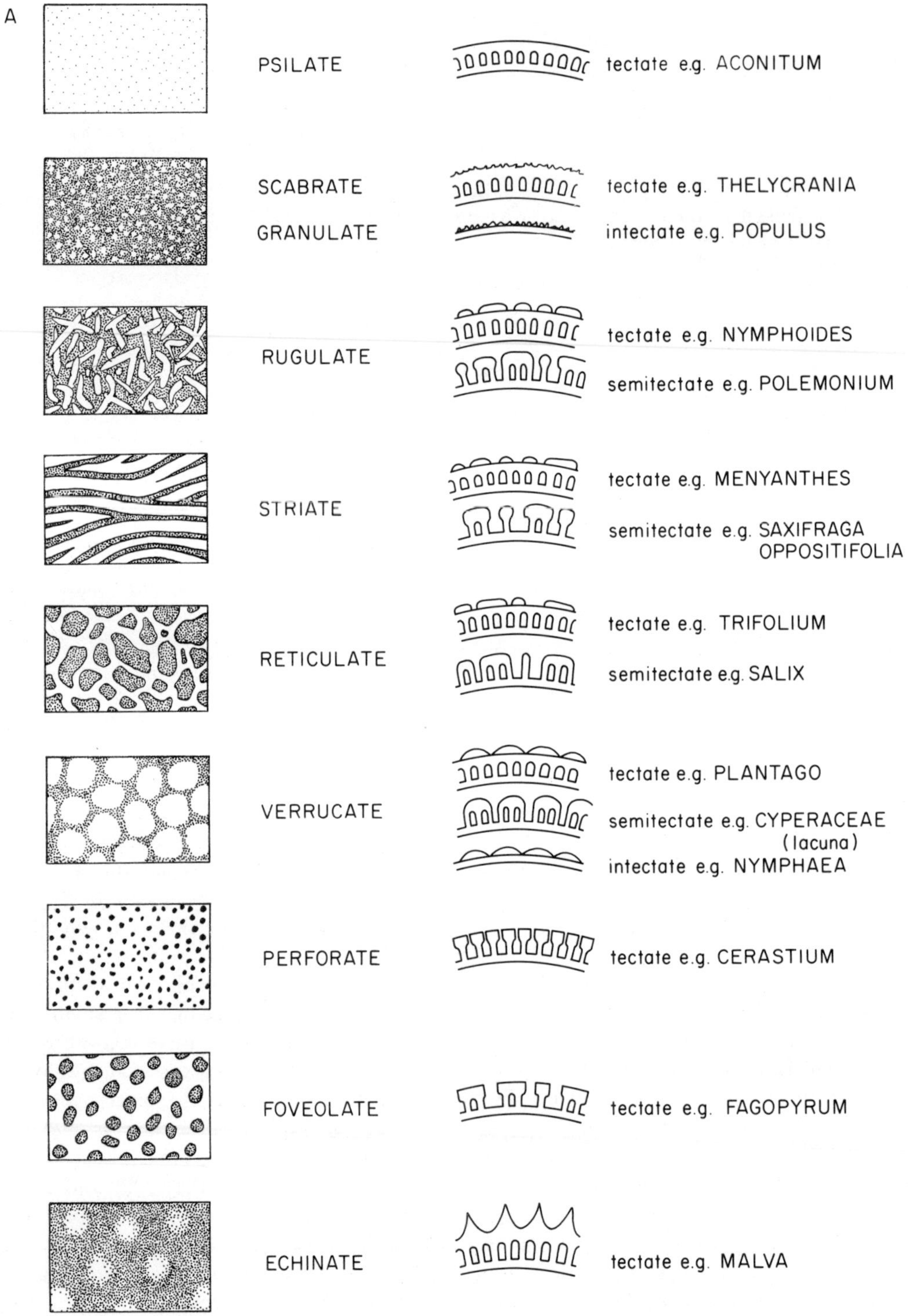

A

PSILATE — tectate e.g. ACONITUM

SCABRATE — tectate e.g. THELYCRANIA

GRANULATE — intectate e.g. POPULUS

RUGULATE — tectate e.g. NYMPHOIDES

semitectate e.g. POLEMONIUM

STRIATE — tectate e.g. MENYANTHES

semitectate e.g. SAXIFRAGA OPPOSITIFOLIA

RETICULATE — tectate e.g. TRIFOLIUM

semitectate e.g. SALIX

VERRUCATE — tectate e.g. PLANTAGO

semitectate e.g. CYPERACEAE (lacuna)

intectate e.g. NYMPHAEA

PERFORATE — tectate e.g. CERASTIUM

FOVEOLATE — tectate e.g. FAGOPYRUM

ECHINATE — tectate e.g. MALVA

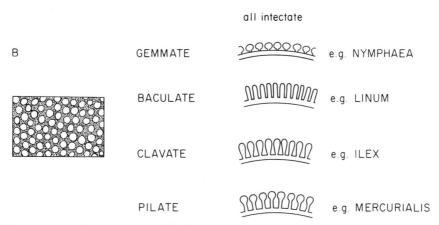

all intectate

B	GEMMATE		e.g. NYMPHAEA
	BACULATE		e.g. LINUM
	CLAVATE		e.g. ILEX
	PILATE		e.g. MERCURIALIS

Fig. 4.8 A. Diagrams of sculpturing types visible in surface view and optical section showing possible underlying exine types. In the sculpturing types all raised areas are shown light, all lower areas or holes are shown dark. It is possible that one sculpturing type, e.g. verrucate, may be produced by three different exine structures. Other sculpturing types, e.g. perforate, can be produced by only one exine structure.

Fig. 4.8 B. Here the same surface pattern is produced by four different sculpturing types. Gemmate, baculate, clavate and pilate all refer to the shape of the projecting processes (see Table 4.1). Theoretically all these types of process could occur on top of the tectum of a tectate grain but no such examples occur within the British flora.

exist which have only a partial tectum, the *semitectate* condition. Examples of semitectate grains are *Salix* or members of the Cruciferae, where the heads of the columellae are joined in two directions only to form winding walls or *muri*. The muri in these cases form a network or *reticulum*. The gaps between the reticulum walls are called *lumina*. Another semitectate pattern is *striate* (e.g. *Acer*) where the muri and lumina run parallel to each other. Patterns that fall between reticulate and striate are called *rugulate*. Illustrations of these types are given in Fig. 4.8.

The number of fine structural variations is theoretically infinite, thus any attempt to segregate pollen grains into neat classes on the basis of fine structure is bound to run into problems. This point can be illustrated easily. A tectum may have perforations of any shape or size in it. If these perforations are large it is often difficult to say whether the grain is then tectate-perforate with large perforations, or semitectate-reticulate with small lumina. If the lumina are equal to or wider than the muri then the grain is described as reticulate and if the muri are wider

than the lumina then the grain is described as tectate-perforate. At the other end of the scale the fenestrate grain of the Compositae, subfamily Liguliflorae, is a type of reticulate grain where the lumina are very large and arranged in a symmetrical pattern. The situation is complicated by the fact that tectate grains may have distinct structures within and upon the tectum. Any structures upon the tectum are described in the same way as free sexinous rods, i.e. according to their shape (e.g. echinae, clavae, bacula, pila, gemmae, verrucae and scabrae). It is also possible for there to be low walls on top of the tectum forming reticulate, striate or rugulate patterns. In tectate grains there is again the problem of types running into one another, for example, *Quercus* falls between scabrate and verrucate and thus has to be designated scabrate-verrucate.

It can be seen from Fig. 4.8 that two grains with the same surface sculpture—say, for example, reticulate, can have very different fine structure. One grain could be tectate with the reticulum walls on top of the tectum and the other could be intectate where the reticulum walls are formed by columellae connected at

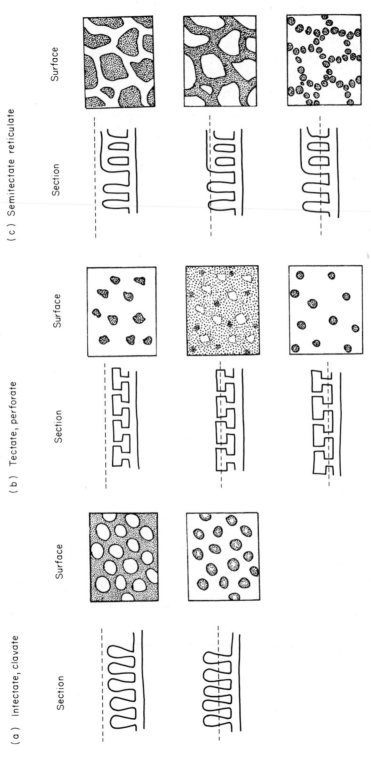

Fig. 4.9 Sculpturing types at different focal planes. The focal plane is indicated by the dashed line on the exine sections. Focusing down through the exine produces diffraction images which vary according to the position of the focal plane. Exine structure may be inferred from the sequence of these images (see 'LO' pattern, page 45).

A. The clavae appear as white islands when plane is above them, dark islands when plane sections them.
B. Perforations appear as dark islands when plane is above the tectum, diffuse white islands when plane is at tectum level. Columellae become distinct dark islands when plane is below the tectum.
C. The reticulum appears as a white network with dark islands (lumina), when plane is above it, dark network with white islands when plane is at the reticulum level. Further downward focusing shows the columellae as distinct dark islands sitting in a reticulate pattern.

their heads. An optical section of the grain would determine which was the case. In the key the two types might be distinguished by calling one tectate, reticulate (or *suprareticulate*) and the other intectate, reticulate (or *eureticulate*).

An optical section does not always make the fine structure of the exine as clear as one might wish, so investigators often deduce a good deal from careful focusing through the sculpturing and patterning presented in a surface view of the grain. The surface types depicted in Fig. 4.8 show any holes or lower areas to be dark and any raised areas or projecting elements to be light. This is how they would appear at highest focus. On focusing carefully down through the exine their appearance would change due to a change in the diffraction images produced. For instance any raised areas or projecting elements might appear dark and any holes, light. Thus a perforation and a columella may both appear as dark spots, but *at different focal planes*. By concentrating attention on a small area of the exine and observing the apparent changes in it as you focus down, the structure of the sexine can be deduced. Fig. 4.9 shows some of the light and dark patterns that result from focusing down through different exine types. This type of investigation was called 'LO'-analysis (from the Latin *lux*, light and *obscuritas*, darkness) by

Erdtman (1956). 'LO' was the term he gave to the sequence: light islands and dark channels (high focus) followed by dark islands and light channels (low focus). If the reverse sequence occurred, i.e. dark islands and light channels followed by light islands and dark channels, it was given the term 'OL'. This system works very well as long as pollen grains and spores are embedded in a medium with a lower refractive index than that of the sporopollenin in the exine.

The use of phase contrast can make the picture of exine fine structure even clearer. Under phase contrast, differences in refractive index are converted to differences in light intensity. Thus the greater the difference in refractive index there is between a solid structure and its surrounding medium, the darker the solid structure will appear to be. This means that very small echinae, scabrae, striae and perforations may become visible where, without phase contrast, they would be indiscernible. Phase contrast also makes easier the detection of a tectum and the investigation of the pattern of the columellae underneath a tectum.

By the use of these features of pollen grain morphology and sculpturing it is possible to identify recent and fossil grains with a fairly high degree of resolution. To this end, the following chapter consists of a key based upon these structural features.

Chapter 5
Pollen and Spore Key with Glossary

The use of the key

To identify any pollen grain or spore it must first be classified according to its aperture type, number and position by the use of the key to pollen classes on page 48. Once assigned to a class, the grains or spores within that class can be separated by reference to sculpturing type, details of aperture structure, etc. Chapter 4 discusses the range of sculpturing and aperture types which may be encountered, and a glossary on page 74 explains the terminology used.

The key is dichotomous, thus at every point of division there are alternatives marked a and b. As far as possible only positive characters have been used, rather than the absence of certain features. Having keyed a grain out to a particular species or type, the relevant photograph will provide confirmation or indicate when a wrong turning has been taken in the key. The photographs should never be used alone for identification as they provide only one (or two) views of a three-dimensional object, and cannot always show the critical features mentioned in the key. This is true particularly where delicate 'LO' analysis (see page 45) with the microscope is necessary to determine exine construction. In some cases, such as when a grain is in a bad condition (crumpled, obscured or eroded) or if it is a member of a difficult morphological group (Rosaceae, Compositae) then identification may prove impossible without type grains for comparison. Ideally all grains identified by means of the key and subsequently compared with the photographs should be checked against type material, since no verbal description or photograph can provide a complete alternative to a type specimen.

When dealing with grains mounted in glycerol jelly a special problem occurs when a grain is presented in an awkward position. With zonocolporate grains an orientation such that one of the poles is uppermost will make the detection of the porus at the equator of each colpus difficult if not impossible. It is possible to change the orientation of individual grains on a slide by pressing gently on the coverslip above the grain with a hot needle. If this proves impossible, taking the grain through both the -colpate and -colporate classes may achieve identification, particularly if the grain has an especially distinctive sculpturing type and is confirmed by the photograph. The detection of a porus in the middle of a colpus may be difficult even if a grain is presented in a good equatorial view. Thickened, circular pori present no problems but unthickened irregular pori are easily missed, and it has been thought that these latter should not be regarded as true pori. In this key any slight constriction, bridge or rupture in the equatorial region of the colpus is taken to represent a porus. Thus most of the Rosaceae are included in trizonocolporate while other authors, e.g. Faegri and Iversen (1964), feel that they should be placed in trizonocolpate.

Spores of some pteridophytes and bryophytes are placed here in the aperture classes which were originally devised to separate angiosperm pollen grains. For instance, pteridophyte spores with one groove are included under monocolpate instead of being placed in a class on their own (*monolete* in other texts such as Erdtman, Praglowski and Nilsson, 1963). Although a groove on a pteridophyte spore is not homologous with an angiosperm colpus, when one is faced with a fossil grain it is often difficult to distinguish the difference. It is hoped that this classification will make some identifications easier. For similar reasons *Ephedra* has been placed in polyzonocolpate rather than in polyplicate.

The use of size measurements has been avoided as much as possible because size varies greatly between grains of the same species, and between grains given different treatments in the extraction process. It should be emphasized that the key is constructed for grains which have been pre-treated with KOH and then acetolysed using the acetic anhydride/sulphuric acid mixture. Acetolysis is known to swell grains to varying extents depending on the duration of the treatment (Reitsma, 1969) and they may be up to 25% larger than grains which have been simply KOH treated. This may not be a problem unless an alternative like 'grain greater than 30 μm or less than 30 μm' occurs in the key. A purely KOH-treated grain might fall below the measurement while an acetolysed grain would fall above it. This would place the KOH-treated grain in the wrong part of the key.

Finally, the debt that this key owes to other major previously published keys, in Faegri and Iversen (1974); McAndrews *et al.* (1973) and Erdtman, Praglowski and Nilsson (1963), must be acknowledged. We hope to have expanded the coverage in certain areas and also to have produced a key which will prove more robust in the hands of the inexperienced, that robustness stemming from the provision of confirmatory illustrations and also from the fact that difficult or variable taxa are keyed out in several places, thus compensating for earlier mistakes. It is hoped that these developments and modifications will encourage undergraduate and sixth-form students to attempt the identification of fossil pollen.

Legends to Plates

Nomenclature follows Clapham, A. R., Tutin, T. G. and Warburg, E. F. (1962) *Flora of the British Isles* (2nd edition), Cambridge University Press.

The pollen type is given first, followed by the identity of the species photographed. Size of the grain is then given in micrometres. In the case of spheroidal grains the diameter only is given, for others the polar diameter is followed by the equatorial diameter of the grain photographed. In the case of tetrads, the full tetrad diameter is given. Where two or three photographs have the same caption, the measurement relates to the first.

Other abbreviations used:

SEM = Scanning electron micrograph.

Ph = Phase contrast micrograph.

Pollen and Spore Key

Key to Pollen Classes

1 a	Grains united in groups:	
	two grains in the group	
	two grains in the group	DYADS (p. 50)
	four grains in the group	TETRADS (p. 50)
	more than four grains in the group	POLYADS (p. 50)
1 b	Grains single	2
2 a	Grains without apertures	3
2 b	Grains with apertures	5
3 a	With sacs or bladders projecting from the body of the grain	SACCATE (p. 50)
3 b	Without such sacs or bladders	4
4 a	With a coarse network of rides separated by lacunae in a fixed geometrical pattern	FENESTRATE (p. 50)
4 b	Without such ridges and lacunae	INAPERTURATE (p. 50)
5 a	With one three-slit aperture, in the shape of a Y	TRILETE (p. 51)
5 b	With $\pm$ circular apertures (pori) or elongate apertures (colpi). An aperture can be combination of colpus + porus	6
6 a	With pori only	7
6 b	With colpi only, or with a porus and a colpus combined in each aperture, or with some apertures colpi, some colpi + pori	9
7 a	With one porus	MONOPORATE (p. 52)
7 b	With more than one porus	8
8 a	Pori arranged in an equatorial zone:	
	two pori in zone	DIZONOPORATE (p. 52)
	three pori in zone	TRIZONOPORATE (p. 53)
	four pori in zone	TETRAZONOPORATE (p. 53)
	five pori in zone	PENTAZONOPORATE (p. 54)
	six pori in zone	HEXAZONOPORATE (p. 54)
8 b	Pori scattered all over the surface of the grain:	
	five pori	PENTAPANTOPORATE (p. 54)
	six pori	HEXAPANTOPORATE (p. 54)
	more than six pori	POLYPANTOPORATE (p. 54)
9 a	With colpi only	
9 b	With a porus and a colpus combined in each aperture or some apertures colpi, some colpi + pori	13
10 a	With one free colpus	MONOCOLPATE (p. 56)
10 b	With more than one colpus. Colpi $\pm$ fused	11
11 a	Colpi fused, often forming rings or spirals	SYNCOLPATE (p. 56)

11 b Colpi free	12
12 a Colpi arranged in an equatorial zone:	
two colpi	DIZONOCOLPATE (p. 57)
three colpi	TRIZONOCOLPATE (p. 58)
four colpi	TETRAZONOCOLPATE (p. 62)
five colpi	PENTAZONOCOLPATE (p. 62)
six colpi	HEXAZONOCOLPATE (p. 62)
more than six colpi	POLYZONOCOLPATE (p. 63)
12 b With colpi scattered over the grain surface:	
four colpi	TETRAPANTOCOLPATE (p. 63)
five colpi	PENTAPANTOCOLPATE (p. 63)
six colpi	HEXAPANTOCOLPATE (p. 63)
more than six colpi	POLYPANTOCOLPATE (p. 63)
13 a Grain with some apertures colpi and some colpi + pori	HETEROCOLPATE (p. 63)
13 b Grain with all apertures colpi + pori	14
14 a Colpi + pori arranged in an equatorial zone:	
three colpi + pori	TRIZONOCOLPORATE (p. 63)
four colpi + pori	TETRAZONOCOLPORATE (p. 72)
five colpi + pori	PENTAZONOCOLPORATE (p. 73)
six colpi + pori	HEXAZONOCOLPORATE (p. 73)
more than six colpi + pori	POLYZONOCOLPORATE (p. 73)
14 b Colpi + pori scattered over the grain surface:	
four colpi + pori	TETRAPANTOCOLPORATE (p. 73)
five colpi + pori	PENTAPANTOCOLPORATE (p. 73)
six colpi + pori	HEXAPANTOCOLPORATE (p. 73)

Dyads (Plate 1)

Grains inaperturate, finely reticulate: *Scheuchzeria*

Tetrads (Plates 1 and 2)

1 a   Tetrad with grains psilate, granulate, scabrate or verrucate: 2
1 b   Tetrad with grains reticulate or echinate: 3
2 a   Each grain trizonoporate, with large vestibulate pori: *Epilobium* type
                              [Includes *Epilobium* and *Oenothera*]
2 b   Each grain either trizonocolpate or trizonocolporate: Ericaceae type
                              [Includes *Empetrum*]
       [See Oldfield, 1959 for further identification of Ericaceae pollen grains.]
3 a   Grains echinate: 4
3 b   Grains reticulate: 6
4 a   With sparse echinae, each 6–8 μm long: *Selaginella*
4 b   With densely packed echinae, each ≤ 4 μm long: 5
5 a   Echinae monomorphic, being of uniform length: *Drosera intermedia*
5 b   Echinae dimorphic, being of two different lengths: *Drosera rotundifolia* type
                              [Includes *D. rotundifolia* and *D. anglica*.]
6 a   Reticulum coarse, muri ≥ 1 μm wide, supported by a single row of coarse columellae. Individual grains without distinct apertures: *Listera* type
                              [Includes *Listera* and *Epipactis*.]
6 b   Reticulum less coarse, muri ≤ 1 μm wide, often supported by more than one row of columellae. Individual grains with distinct apertures: *Typha latifolia* type

Polyads

Individual grains of group not regularly arranged and so pressed together that outline of individual grains becomes angular: Orchidaceae

Polyads are also found in the Mimosaceae but here the grains are very regularly arranged into a convex disc (see Fig. 4.4) and individual grains tend to retain their circular outline.

Saccate (Plate 2)

1 a   Sacci not constricted at the point of attachment to the body of the grain. Area of sac attachment large, distance between sacs smaller: *Picea*

1 b   Sacci constricted at the point of attachment to the body of the grain. Area of sac attachment small, distance between sacs larger: 2
2 a   Body of grain in size range 40–60 μm: *Pinus*
2 b   Body of grain in size range 70–100 μm: *Abies*

Fenestrate (Plate 2)

Grains with a coarse network of high echinate ridges separated by large spaces (lacunae): Compositae, Liguliflorae
       [Includes *Taraxacum, Tragopogon, Arnoseris, Cichorium, Crepis, Hieracium, Hypochaeris, Lapsana, Leontodon, Lactuca, Picris, Sonchus, Cicerbita.*]

Although not strictly an aperture system, fenestrate (which means 'windows') is here used because the high echinate ridges obscure the true aperture system of these grains. The true aperture system can be seen as either trizonoporate or trizonocolporate. See the diagrams in Wodehouse (1935).

Inaperturate (Plates 3, 4 and 5)

1 a   With echinae, or psilate, or possessing two coats, the outer one variously wrinkled and folded: 2
1 b   Without echinae or wrinkled outer coat. Grains may be granulate, gemmate, clavate or reticulate: 10
2 a   Echinate: 3
2 b   Smooth or possessing two coats: 6
3 a   Echinae > 5 μm long, no columellae visible in the area between the echinae: *Nuphar*
                              [See also *Cystopteris* and *Thelypteris*, page 57.]
3 b   Echinae < 5 μm long, columellae ± visible between them: 4
4 a   Echinae cylindrical, often arranged at the intersections of a reticulum. Columellae visible in the area between the echinae (× 1000): *Stratiotes*
4 b   Grain > 35 μm, ± spherical: 5
5 a   Echinae approximately 1·5 μm long, cylindrical: *Hydrocharis*
5 b   Echinae > 1·5 μm long, triangular: *Lemna*
                              [See also page 52.]
6 a   Grains without wrinkled additional outer coat (i.e. smooth): 7
6 b   Grains with a variously wrinkled additional outer coat, ± smooth: 8
7 a   Grains usually spherical, but often split, > 40 μm: *Larix*
7 b   Grains bean shaped, usually < 40 μm: Polypodiaceae

8 a   Grains spherical: *Equisetum*

8 b   Grains bean shaped (shape produced mainly by inner coat): 9

9 a   With few folds or wrinkles, pattern lacking: *Isoetes*

                [See also *Blechnum*, page 57.]

9 b   With outer coat distinctly folded, wrinkled or ridged: Polypodiaceae

        [For further differentiation see page 57.]

10 a Grains reticulate, reticulum formed either by the joined heads of the columellae or by separate columellae sitting in a reticulate pattern: 11

10 b Grains with structural elements either randomly arranged or in a dense, $\pm$ uniform carpet. Structural elements can be gemmae, bacula and clavae as well as columellae: 14

11 a Heads of columellae joined (black network in phase): 12

11 b Heads of columellae free (isolated black dots in phase): 13

12 a Exine thin (< 2 μm). Reticulum fine and lace-like: *Potamogeton* section *Eupotamogeton*

       [Includes some species of *Potamogeton* and *Triglochin*.]

12 b Exine thicker (> 3 μm), columellae long (3 μm). Reticulum coarse: *Daphne*

                [See also page 56.]

13 a Grain with clavae. Exine > 1 μm thick. Grains up to 20 μm: *Callitriche*

13 b Grain with bacula. Exine approximately 1 μm thick. Grains usually > 20 μm: *Potamogeton* section *Coleogeton*

14 a With most of surface tectate, perforate (in × 1000 and phase), but with areas (lacunae) where the surface is semitectate, verrucate. Grains pear-shaped or roughly isodiametric, often collapsed or crumpled: Cyperaceae

14 b Grains intectate, no lacunae, grains $\pm$ isodiametric: 15

15 a With at least some structural elements long and swollen headed (i.e. clavae): 16

15 b Elements gemmae, microgemmae or minute granules (there may be a mixture of all three, but no clavae present): 17

16 a Elements as a whole varying in height and thickness (i.e. having mixed clavae, bacula, gemmae and verrucae): *Nymphaea*

      [See also page 53, *Athyrium* (page 57) may also key out here if the furrow is not visible.]

16 b Elements all the same height and all clavae: *Callitriche*

17 a Elements all microgemmae, appearing in very irregularly scattered groups (use phase contrast if possible). There may be very few gemmae left

because they are deciduous. Grains often split (see Plate 4): *Juniperus* type

           [Includes *Juniperus* and *Cupressus*.]

17 b Structural elements gemmae or minute granules, much more crowded and in a dense $\pm$ uniform carpet: 18

18 a Grain spherical with minute granules which are only really visible in phase contrast. Granules varying in width and height: *Populus tremula*

18 b Grain angular (sometimes rectangular-obtuse). Elements bigger and resembling gemmae. These gemmae may vary in size. Frequently splitting, cf. *Juniperus*: *Taxus*

      [Grains resembling those of *Taxus* and *Populus* may be found within many moss species, e.g. *Polytrichum* and *Funaria* (see Dickson, 1973). Erdtman (1969) reports that some algal cysts have been noticed to have this sculpturing type.]

Trilete (Plates 5, 6 and 7)

1 a   Spores either reticulate, foveolate or with irregular creases: 2

1 b   Spores psilate, granulate, echinate or verrucate-undulate: 9

2 a   Surface with regularly spaced foveolae. Spore shape triangular with the angles trunkated: *Lycopodium selago*

2 b   Without foveolae but with a thin-walled reticulum or irregular creases: 3

3 a   With an outer coat thrown into irregular creases on the distal surface of the spore and small rounded verrucae on the proximal surface: 4

3 b   With a reticulum on the distal surface: 5

4 a   Spores > 40 μm, creases large and strong: *Lycopodium inundatum*

4 b   Spores < 40 μm, creases weaker: *Sphagnum*

5 a   Muri varying markedly in thickness and in height: *Ophioglossum*

5 b   Muri all the same thickness and height: 6

6 a   Adjacent lumina in the centre of the distal face varying markedly in size and shape—some small and isodiametric, others large, elongate and winding. Largest lumen always > 9 μm in width (or length): *Lycopodium alpinum*

6 b   Adjacent lumina in the centre of the distal face not varying markedly in size and shape, being all roughly isodiametric. Largest lumen may be greater or less than 9 μm: 7

7 a   Reticulum appearing large and coarse. Largest lumen $\geqslant$ 9 μm in width. Ratio of spore size to number of lumina across distal face usually > 3·0: *Lycopodium annotinum*

7 b   Reticulum not so coarse. Largest lumen < 9 μm in width, ratio of spore size to number of lumina across distal face usually < 3·0: *Lycopodium clavatum*

8 a   Psilate or granulate, but not with verrucae or echinae: 9

8 b   Echinate or verrucate: 12

9 a   Spores > 50 μm in width, circular in shape: 10

9 b   Spore < 50 μm in width, circular to triangular obtuse in shape: 11

10 a   Psilate, spore wall up to 13 μm thick and visibly two-layered: *Pilularia*

10 b   Psilate to granulate, spore wall < 5 μm thick, not visibly two-layered: *Osmunda*

11 a   With either irregular granulate-gemmate sculpturing, or completely psilate. Spore wall beneath granules or gemmae the same thickness all over and uniformly thin. Arms of three-slit aperture not reaching the outline of the spore in proximal view, thus proximal face is not angled. Shape of spore tending towards triangular-obtuse-concave: *Pteridium*

11 b   Spores ± granular-gemmate, but commonly psilate. Spore wall beneath granules or gemmae tending to be thickest at the corners of the triangle. Arms of the three-slit aperture nearly reaching the outline of the spore in proximal view, thus proximal face strongly angled. Shape of spore tending towards triangular-obtuse-convex: *Sphagnum*

12 a   Spores with sparse echinae (6–8 μm long) on the distal surface: *Selaginella*

12 b   Spores without echinae, any projecting processes < 5 μm long: 13

13 a   Spores > 50 μm, proximal face not angled between the arms of the three-slit aperture: 14

13 b   Spores < 50 μm, proximal face ± angled between the arms of the three-slit aperture: 15

14 a   Spore circular to triangular-obtuse, with large undulating verrucae: *Cryptogramma*

14 b   Spore circular but never triangular-obtuse, surface covered in a mixture of small irregular gemmae and verrucae: *Osmunda*

15 a   With verrucae or large undulating rugulae: 16

15 b   With non-hemispherical, commonly elongate and interconnecting processes: 17

16 a   Shape tending towards triangular-obtuse-concave, verracae commonly elongated into rugulae, arms of three-slit aperture not reaching spore outline in proximal view: *Botrychium*

16 b   Shape triangular-obtuse-convex, verrucae not elongated, arms of three-slit aperture commonly reaching spore outline in proximal view, proximal face sharply angled: *Sphagnum*

17 a   Spore shape tending towards circular, spore wall ⩾ 3 μm thick. Arms of three-slit aperture not reaching spore outline in proximal view, sculpturing very coarse: *Ophioglossum*

17 b   Spore shape tending towards triangular-obtuse-convex. Spore wall up to three μm. Arms of three-slit aperture commonly reaching spore outline in proximal view, proximal face therefore sharply angled: *Sphagnum*.

N.B. For some good scanning electron micrographs of trilete spores see Moe (1974).

Monoporate (Plates 7 and 8)

1 a   Porus large, diameter more than half that of grain. Porus operculate. Surface of grain with mixed clavae, gemmae, verrucae and bacula: *Nymphaea*

1 b   Porus smaller, diameter less than half that of grain: 2

2 a   Porus well defined and circular. Nexine thickened around it to form a costa. Grains tectate with verrucate, rugulate scabrate or microverrucate sculpturing. Columellae ± visible (× 1000, phase). Columellae may be evenly distributed or aggregated: Graminae

2 b   Porus not so well defined and without thickened edge. Porus edge may be ragged and porus itself difficult to locate: 3

3 a   Grain tectate perforate (in phase, × 1000), columellae visible and uniformly distributed under tectum. Sometimes areas that are semitectate, verrucate are visible on the sides of the grain. Grains often pear-shaped or crumpled: Cyperaceae

3 b   Grains intectate, either echinate or eu-reticulate: 4

4 a   Echinate, projections broad based (almost conical, 1·5 μm long). Porus very obscure: *Lemna*

4 b   Reticulate, with lumina of differing sizes. Porus defined only by absence of reticulate pattern: *Typha augustifolia* type
         [Includes *T. augustifolia* and *Sparganium*.]

Dizonoporate

Colchicum is the only grain which normally has two pori, however it does not survive fossilization as the grains are very thin walled and explode on wetting. Dizonoporate (and Dipantoporate) grains occasionally occur in genera with normally three or more porate grains, e.g. *Betula*, *Myriophyllum*.

Trizonoporate (Plates 8, 9 and 10)

1 a   Grains > 50 μm: 2
1 b   Grains < 50 μm: 4
2 a   With vestibulate pori, vestibulum large and cylindrical. Exine without echinae or reticulations: *Epilobium* type
     [Includes *Epilobium*, *Circaea* and *Oenothera*.]
2 b   Without vestibulate pori, grains either echinate or reticulate: 3
3 a   Columellae branching and anastomosing to form a reticulum. Clavae-like projections on top of the muri: *Geranium* type
     [See also trizonocolporate, reticulate, page 66. Includes *Geranium* and *Erodium*.]
3 b   Exine covered with a dense carpet of small echinae and sparsely spaced larger echinae (up to 3 μm long): *Knautia* type
     [Includes *Knautia* and *Dipsacus*.]
4 a   Grains with protruding, vestibulate pori: 5
4 b   Grains with $\pm$ protruding, non-vestibulate pori: 6
5 a   Vestibulum longer than wide, with straight sides (i.e. cylindrical). Grain > 30 μm: *Circaea*
     [See *Epilobium* type above.]
5 b   Vestibulum wider than long, with slanting sides (see Fig. 4.6). Most grains < 30 μm: *Betula*
     [Includes *B. pubescens*, *B. pendula* and *B. nana*.]
6 a   Grain circular in polar view, pori sunken or very slightly protruding: 7
6 b   Grain circular to triangular-obtuse in polar view, pori situated on the angles, $\pm$ protruding: 10
7 a   Pori slightly protruding due to thickening of the exine in the porus area. Grains not echinate: 8
7 b   Pori flat or sunken but not protruding. Grains $\pm$ echinate: 9
8 a   Exine becoming suddenly thickened at the porus margin. Hollow part of porus characteristically U-shaped (see Plate 9). Exine away from the vicinity of the pori is very thin ($\sim$ 0·5 μm): *Cannabis* type
     [Includes *Cannabis* and *Humulus*.]
8 b   Exine becoming gradually thickened towards the porus margin. Hollow part of the porus not characteristically U-shaped. Exine away from the vicinity of the pori > 0·5 μm thick: *Corylus*

N.B. *Betula nana* may key out here because in some grains the vestibulum is very inconspicuous.

9 a   Grains with echinae, pori large and in sunken

portions of the exine. Exine > 0·5μm thick: *Campanula* type
     [Includes *Campanula*, *Phyteuma*, *Wahlenbergia* and *Jasione*.]
9 b   Grains without echinae. Pori small and with a little thickening at the margins, but pori never in sunken areas. Exine < 0·5 μm thick: *Urtica* type
     [Includes *Urtica* and *Parietaria*.]
10 a   Nexine absent in the vicinity of the pori thus the 'inside' diameter of the porus opening is greater than the 'outside' diameter of the porus opening (see Fig. 4.6). $\times$ 1000 magnification and phase contrast shows the sculpturing to consist of microechinae (appearing as small dots) on top of the tectum: *Myrica*
10 b   Nexine may be split into a few lamellae near the porus, but never ceases suddenly, thus porus external diameter is much the same as the internal diameter. Under $\times$ 1000 and phase contrast the sculpturing appears faintly rugulate, with microechinae on the ridges: *Corylus*

N.B. *Betula nana* may key out here if vestibulum is insignificant. See Erdtman, Praglowski and Nilsson (1963) for differentiation of *Betula*, *Corylus* and *Myrica* on porus structure.

Tetrazonoporate (Plates 10 and 11)

1 a   Grains echinate or with broad undulating rugulae or reticulum: 2
1 b   Grains psilate, scabrate, microechinate or microrugulate: 3
2 a   Grains echinate, with pori in slightly sunken areas of the exine: *Campanula* type
     [Includes *Campanula*, *Phyteuma*, *Wahlenbergia* and *Jasione*.]
2 b   Grain with broad winding ridges or shallow, wide-walled reticulum. Pori not sunken, but very slightly protruding: *Ulmus*
3 a   Pori vestibulate, neighbouring pori connected by arcs or bands of nexinous thickening: *Alnus*
3 b   Pori non-vestibulate, $\pm$ thickened. If pori thickened then thickenings always solid: 4
4 a   Grains < 20 μm in size: 5
4 b   Grains > 20 μm in size: 6
5 a   Exine approximately 0·5 μm thick (slightly thickened around the pori). Pori very small in comparison to grain size. Grain surface may appear microechinate: *Urtica* type
     [*Urtica* and *Parietaria*.]
5 b   Exine > 0·5 μm thick (may be thinner around pori). Pori size up to 25% of the diameter of the grain. Grain surface scabrate: *Herniaria* type
     [Includes *Herniaria* and *Illecebrum*.]

6 a   Pori margins protruding slightly, but nexine in this marginal area dissolved so that exine near pori ⩽ the thickness of the exine elsewhere. Sculpture approaches microrugulate: *Carpinus*

6 b   Pori ± protruding, nexine never dissolved at the porus margin, therefore exine near pori ⩾ the thickness of the exine elsewhere: 7

7 a   Pori asymmetrically distributed around the circumference of the grain, pori margins markedly thickened and protruding (margin > 3 μm thick): *Myriophyllum alterniflorum*

7 b   Pori symmetrically distributed around the circumference of the grain, pori margins ⩽3 μm in thickness. Porus shape elliptic rather than circular: 8

8 a   Pori margins slightly protruding, approximately 3 μm thick: *Myriophyllum spicatum*

8 b   Pori margins not protruding, approximately 2–5 μm thick: *Myriophyllum verticillatum*

Pentazonoporate, Hexazonoporate (Plate 11)

1 a   Pori vestibulate. Grain with arcs or bands of nexinous thickening joining adjacent pori: *Alnus*

1 b   Pori non-vestibulate, with margins either unthickened or solidly thickened. No arcs of thickening travelling between adjacent pori: 2

2 a   Grain with broad rugulae or a shallow, wide-walled reticulum covering the surface. Tectum undulating in optical section of grain wall. Nexine not dissolved near the pori: *Ulmus*

2 b   Grains scabrate or microrugulate, pori protruding but margins are unthickened. Nexine in marginal area slightly dissolved: *Carpinus*

Pentapantoporate, Hexapantoporate (Plates 11 and 12)

1 a   Each porus with an annulus or costa (i.e. a difference in the thickness of sexine or nexine near the porus). Pori always circular with edges well defined, ± operculate: 2

1 b   No such differentiated area around each porus, edge of porus either well defined or ragged and diffuse, porus membrane granulate: 5

2 a   Pori not protruding in an optical section of the grain, exine thinner at the porus edge than in the middle of the interporium. Tectum perforate to eureticulate: Caryophyllaceae
       [For differentiation of types within this family see Faegri and Iversen, 1974; Chanda, 1962.]

2 b   Pori protruding slightly in an optical section of the grain. Exine ± thicker at the porus margins: 3

3 a   Nexine slightly dissolved in the region around the pori, exine thus thinner in this area or the same thickness as elsewhere. Grains scabrate or microrugulate (with small echinae on the ridges): *Carpinus*

3 b   Nexine not dissolved near the pori, exine always thicker here than in the interporium. Thus annulus appears as a solid ring around each porus. Grain sculpturing scabrate or with large verrucae: 4

4 a   Grain with six large pori—diameter of one porus + annulus > 25% the diameter of the grain. Pori not operculate: *Fumaria*

4 b   Grains with smaller pori—diameter of one porus + annulus is ⩽25% of the diameter of the grain. Pori commonly operculate: *Plantago lanceolata* type
       [Includes *P. lanceolata* and *P. coronopus*.]

5 a   Pori situated in slightly sunken areas of the exine. Porus margin very diffuse and porus membrane granulate (often pori are difficult to distinguish from the mesoporium). Grains scabrate or microechinate, but never with large verrucae: *Thalictrum* type
       [Includes *T. flavum*, *T. minus* and *T. alpinum*.]

5 b   Pori not in slightly sunken areas but flush with the grain surface. Grains verrucate, microechinate or a combination of the two: 6

6 a   Pori sharply delimited. Sculpturing ± verrucate, but always with distinct regularly spaced microechinae. Columellae visible in phase contrast as a dense carpet underneath the verrucae and microechinae: *Plantago maritima*

6 b   Pori not sharply delimited (presence is shown only by the cessation of verrucate sculpturing). Sculpturing always verrucate, columellae fine to invisible. Microechinae ± visible: *Plantago media/major*
       [A distinction between *P. major* and *P. media* may be made by comparison with type slides. The distinction depends on the visibility of the microechinae.]

Polypantoporate (Plates 12 to 16)

1 a   Grains with surfaces psilate-scabrate to microechinate: 2

1 b   Grains with surfaces either echinate, verrucate, reticulate, or rugulate-striate, but never psilate-scabrate: 10

2 a   Each porus with a differentiated area around it. This area may have either sexine or nexine thicker or thinner than the rest of the mesoporium. Pori always circular with the edges well defined: 3

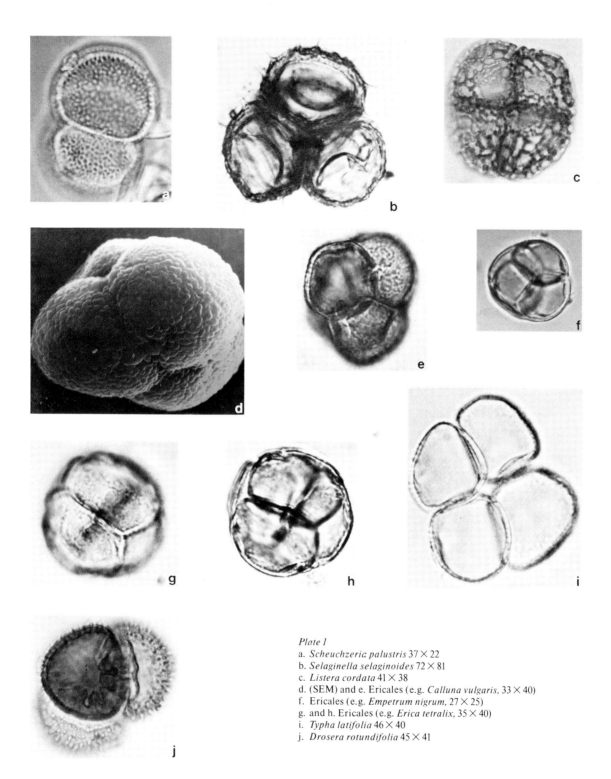

Plate 1
a. *Scheuchzeria palustris* 37 × 22
b. *Selaginella selaginoides* 72 × 81
c. *Listera cordata* 41 × 38
d. (SEM) and e. Ericales (e.g. *Calluna vulgaris*, 33 × 40)
f. Ericales (e.g. *Empetrum nigrum*, 27 × 25)
g. and h. Ericales (e.g. *Erica tetralix*, 35 × 40)
i. *Typha latifolia* 46 × 40
j. *Drosera rotundifolia* 45 × 41

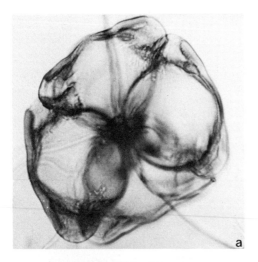

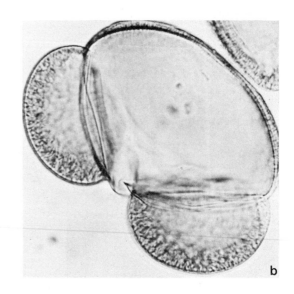

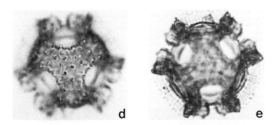

Plate 2
a. *Epilobium* type (e.g. *E. hirsutum*, 140 × 118)
b. *Abies* (e.g. *A. alba*, 107 × 65)
c. *Picea* (e.g. *P. glauca*, 99 × 75)
d. and e. Compositae, Liguliflorae (e.g. *Taraxacum officinale*, 41 × 38)
f. and g. *Pinus* (e.g. *P. sylvestris*, 57 × 37)

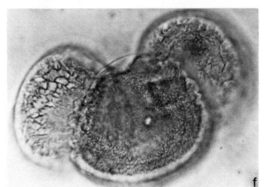

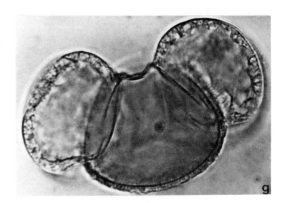

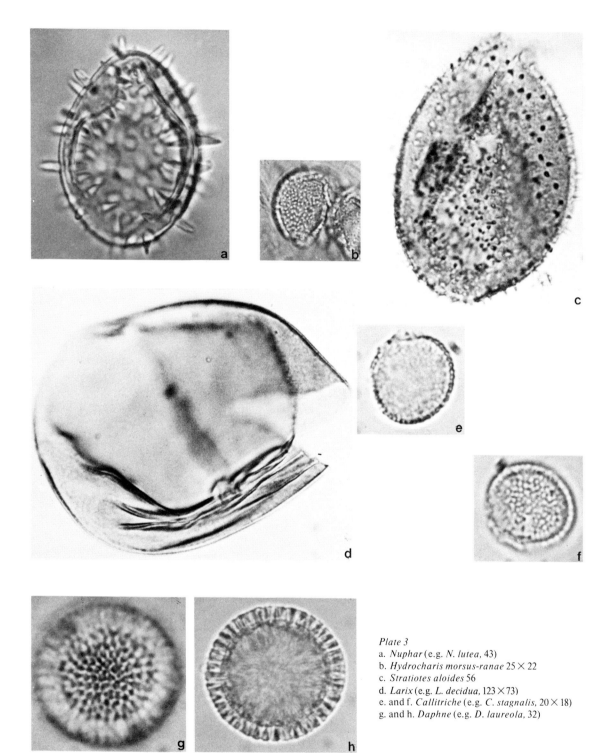

Plate 3
a. *Nuphar* (e.g. *N. lutea*, 43)
b. *Hydrocharis morsus-ranae* 25 × 22
c. *Stratiotes aloides* 56
d. *Larix* (e.g. *L. decidua*, 123 × 73)
e. and f. *Callitriche* (e.g. *C. stagnalis*, 20 × 18)
g. and h. *Daphne* (e.g. *D. laureola*, 32)

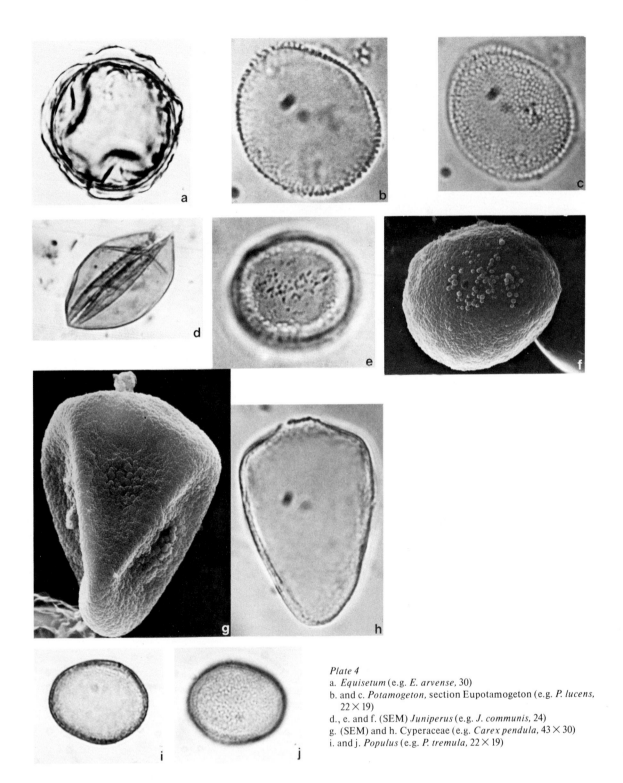

Plate 4

a. *Equisetum* (e.g. *E. arvense*, 30)

b. and c. *Potamogeton*, section Eupotamogeton (e.g. *P. lucens*, 22 × 19)

d., e. and f. (SEM) *Juniperus* (e.g. *J. communis*, 24)

g. (SEM) and h. Cyperaceae (e.g. *Carex pendula*, 43 × 30)

i. and j. *Populus* (e.g. *P. tremula*, 22 × 19)

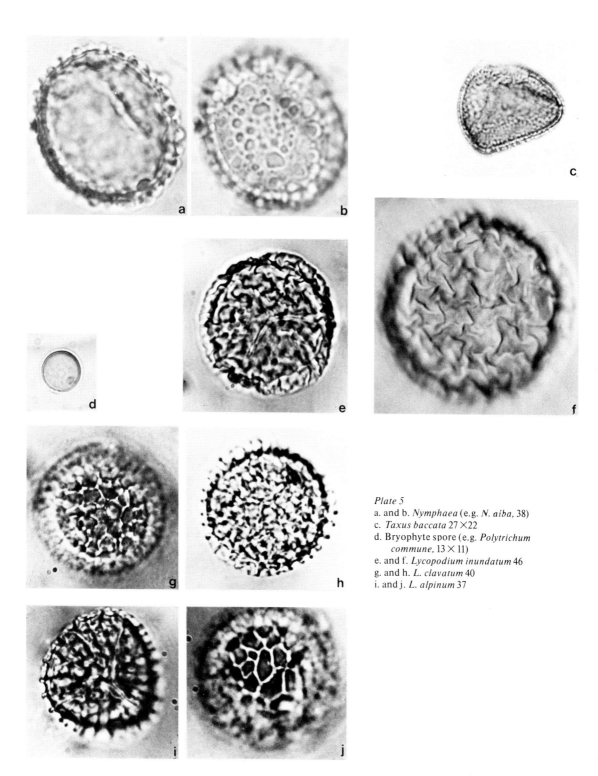

Plate 5
a. and b. *Nymphaea* (e.g. *N. aíba*, 38)
c. *Taxus baccata* 27 ×22
d. Bryophyte spore (e.g. *Polytrichum*
      commune, 13 × 11)
e. and f. *Lycopodium inundatum* 46
g. and h. *L. clavatum* 40
i. and j. *L. alpinum* 37

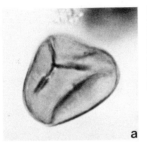

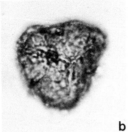

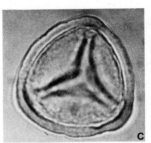

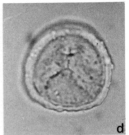

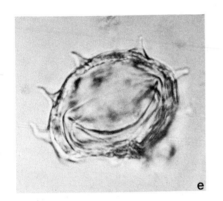

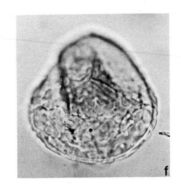

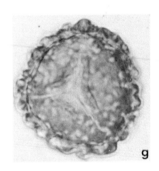

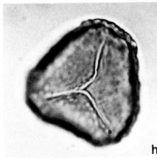

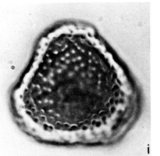

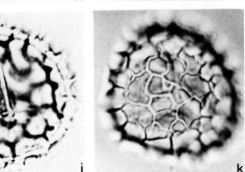

Plate 6
a. and b. *Pteridium aquilinum* 35
c. *Sphagnum* (e.g. *S. tenellum*, 37)
d. *Sphagnum* (e.g. *S. papillosum*, 33)
e. *Selaginella selaginoides* 41 × 37
f. *Botrychium lunaria* 37
g. *Ophioglossum vulgatum* 41
h. and i. *Lycopodium selago* 41
j. and k. *L. annotinum* 40

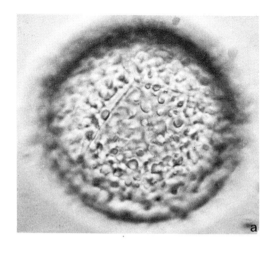

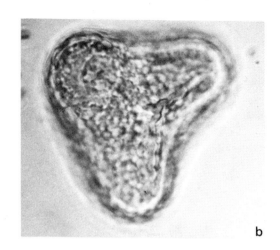

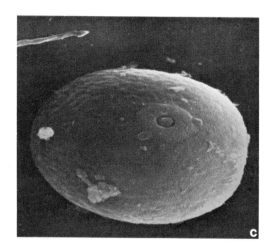

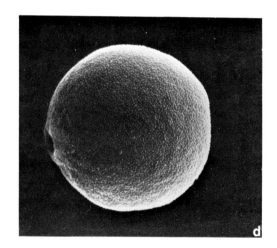

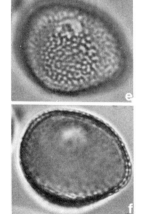

Plate 7
a. *Osmunda regalis* 72
b. *Cryptogramma crispa* 76
c. (SEM) Gramineae (e.g. *Secale cereale,* 49 × 35)
d. (SEM) Gramineae (e.g. *Aira* sp., 30)
e. and f. *Typha angustifolia* type (e.g. *T. angustifolia,* 25)

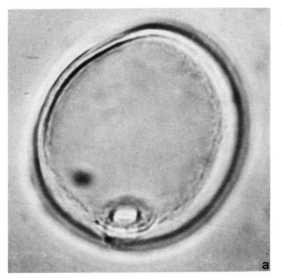

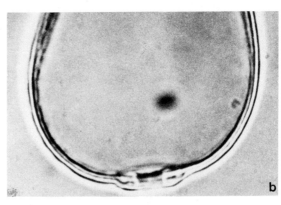

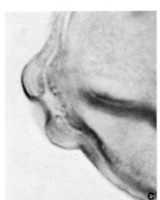

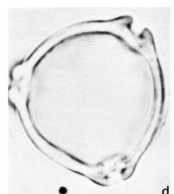

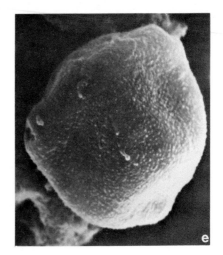

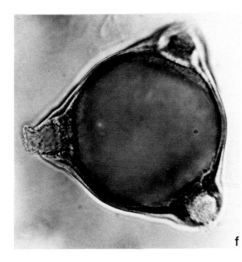

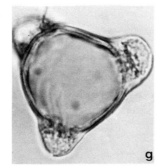

Plate 8
a. Gramineae (e.g. *Triticum turgidum*, 46)
b. Gramineae (e.g. *Avena sativa*. 59 × 41)
c., d. and e. (SEM) *Betula* (e.g. *B. pendula*, 35)
f. *Epilobium* type (e.g. *E. angustifolium*, 71)
g. *Circaea lutetiana* 40

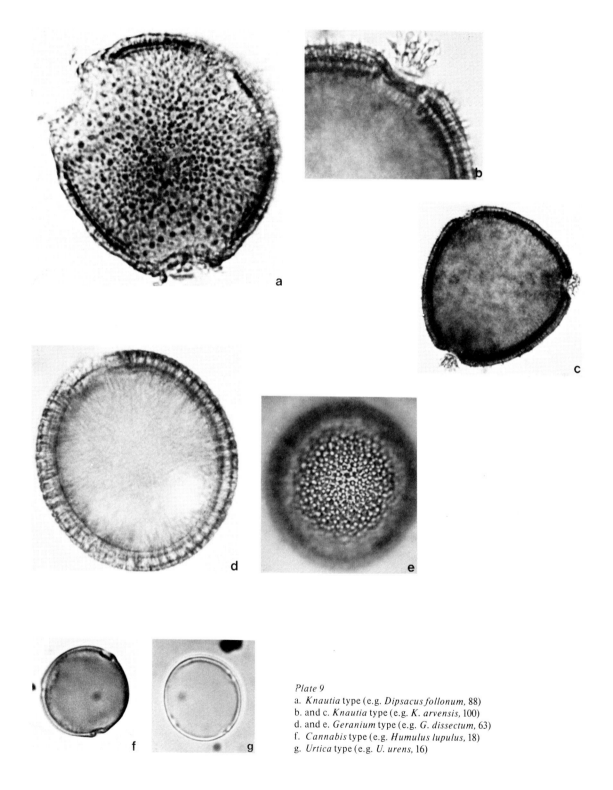

Plate 9
a. *Knautia* type (e.g. *Dipsacus follonum*, 88)
b. and c. *Knautia* type (e.g. *K. arvensis*, 100)
d. and e. *Geranium* type (e.g. *G. dissectum*, 63)
f. *Cannabis* type (e.g. *Humulus lupulus*, 18)
g. *Urtica* type (e.g. *U. urens*, 16)

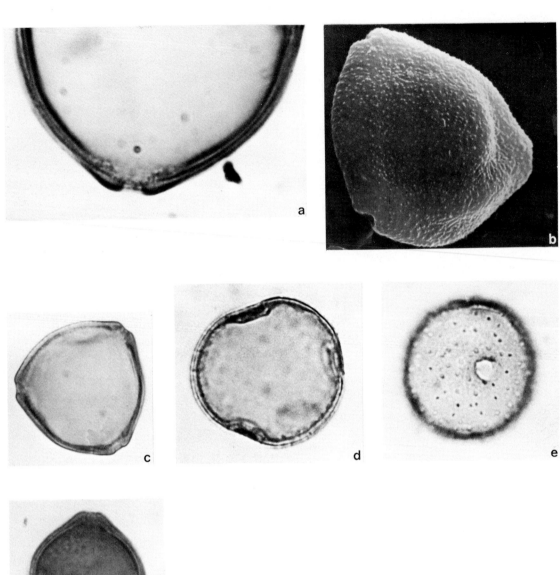

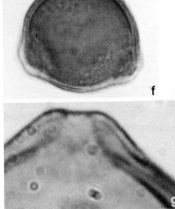

Plate 10
a., b. (SEM) and c. *Corylus avellana* 43
d. and e. *Campanula* type (e.g. *C. rotundifolia*, 37)
f. and g. *Myrica gale* 38

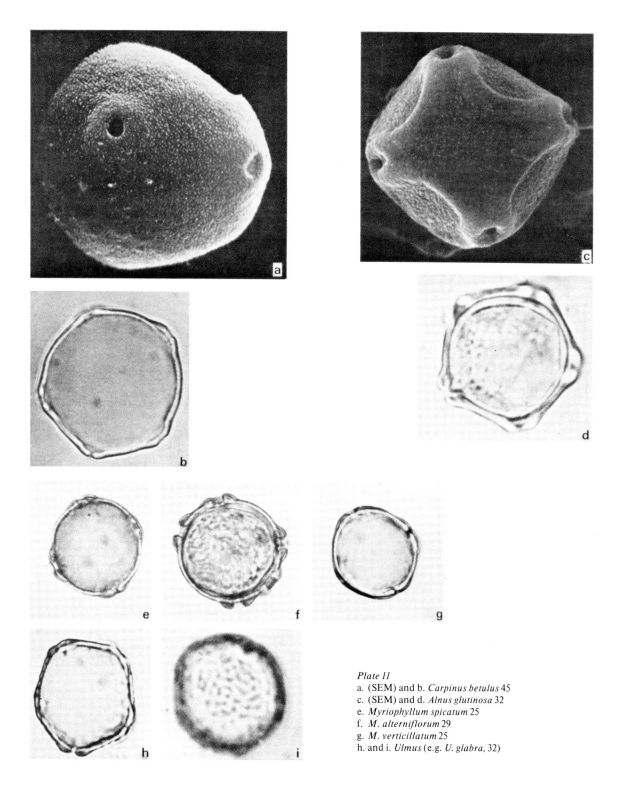

Plate 11
a. (SEM) and b. *Carpinus betulus* 45
c. (SEM) and d. *Alnus glutinosa* 32
e. *Myriophyllum spicatum* 25
f. *M. alterniflorum* 29
g. *M. verticillatum* 25
h. and i. *Ulmus* (e.g. *U. glabra*, 32)

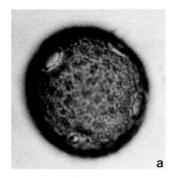
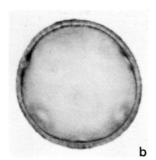
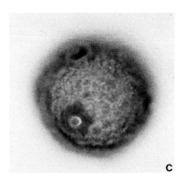
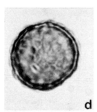
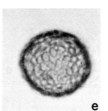
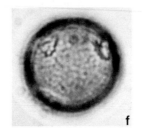
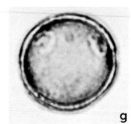
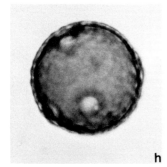
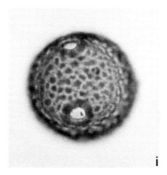
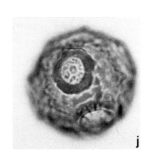

Plate 12
a. *Plantago maritima* 32
b. and c. *P. lanceolata* 32
d. and e. *Plantago media/major* type (e.g. *P. media*, 22)
f. and g. *Thalictrum* (e.g. *T. alpinum*, 18)
h. and i. *Plantago coronopus* 37
j. *Fumaria* (e.g. *F. officinalis*, 33)

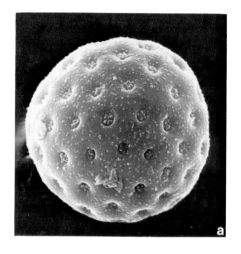

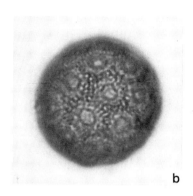

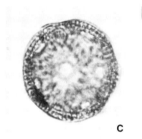
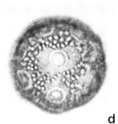
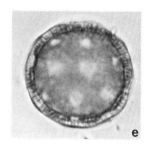
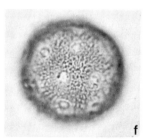

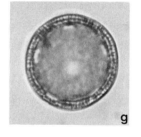
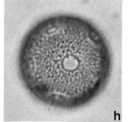
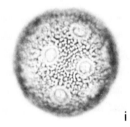

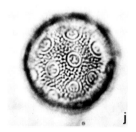
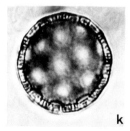

Plate 13
a. (SEM) Chenopodiaceae (e.g. *Chenopodium* sp., 25)
b. Chenopodiaceae (e.g. *Salsola kali*, 30)
c. and d. Caryophyllaceae (e.g. *Scleranthus annuus*, 35)
e. and f. Caryophyllaceae (e.g. *Cerastium alpinum*, 37)
g. and h. Caryophyllaceae (e.g. *Myosoton aquaticum*, 32)
i. Caryophyllaceae (e.g. *Silene vulgaris*, 40)
j. and k. Caryophyllaceae (e.g. *Lychnis flos-cuculi*, 32)

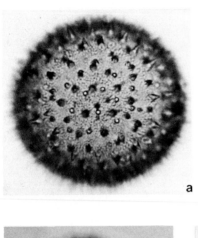

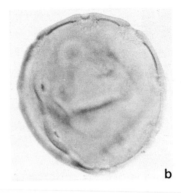

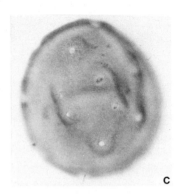

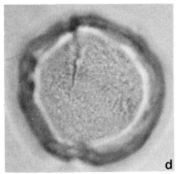

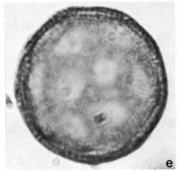

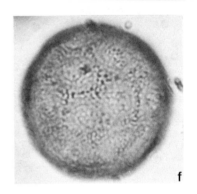

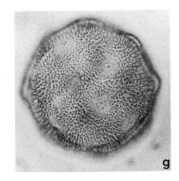

Plate 14
a. Malvaceae (e.g. *Malva sylvestris*, 102)
b. and c. *Juglans regia* 40
d. *Trientalis europaea* (polypantoporate grain) 29
e. and f. *Alisma* (e.g. *A. plantago-aquatica*, 22)
g. *Calystegia* (e.g. *C. sepium*, 71)
h. and i. *Sagittaria* (e.g. *S. sagittifolia*, 24)

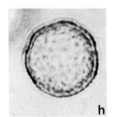

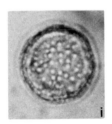

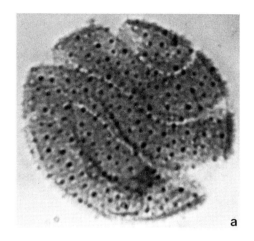

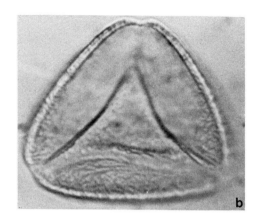

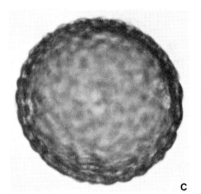

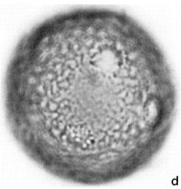

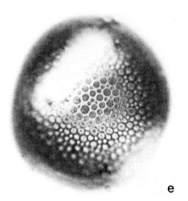

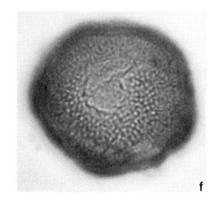

Plate 15
a. *Eriocaulon septangulare* 25
b. *Nymphoides peltata* 35
c. and d. *Littorella uniflora* 52
e. *Linum anglicum* 78
f. *Papaver argemone* 40

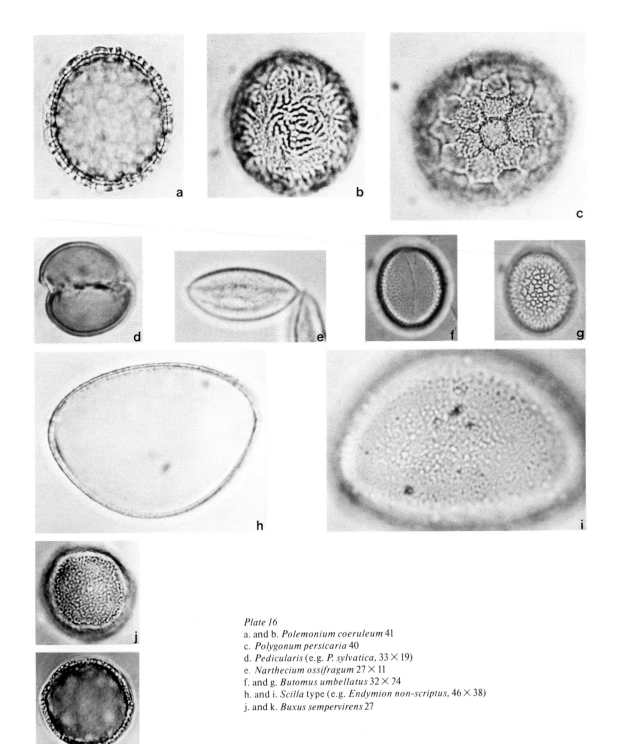

Plate 16
a. and b. *Polemonium coeruleum* 41
c. *Polygonum persicaria* 40
d. *Pedicularis* (e.g. *P. sylvatica*, 33 × 19)
e. *Narthecium ossifragum* 27 × 11
f. and g. *Butomus umbellatus* 32 × 24
h. and i. *Scilla* type (e.g. *Endymion non-scriptus*, 46 × 38)
j. and k. *Buxus sempervirens* 27

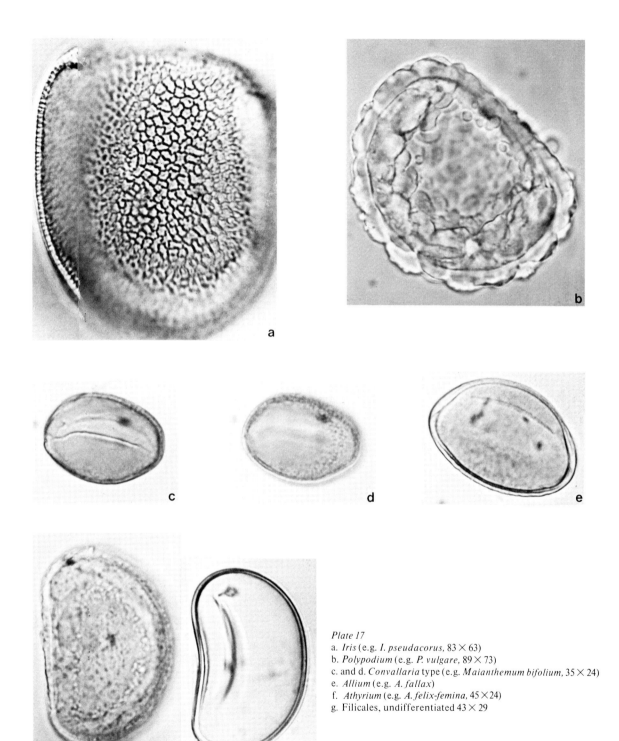

Plate 17
a. *Iris* (e.g. *I. pseudacorus*, 83 × 63)
b. *Polypodium* (e.g. *P. vulgare*, 89 × 73)
c. and d. *Convallaria* type (e.g. *Maianthemum bifolium*, 35 × 24)
e. *Allium* (e.g. *A. fallax*)
f. *Athyrium* (e.g. *A. felix-femina*, 45 × 24)
g. Filicales, undifferentiated 43 × 29

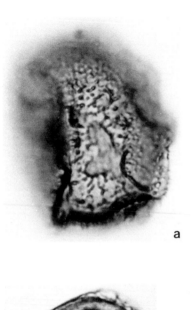

a

b

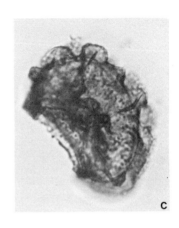

c

d

e

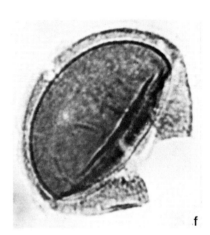

f

g

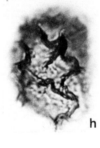

h

Plate 18
a. *Dryopteris dilatata* 49 × 41
b. *Dryopteris filix-mas* type (e.g. *D. aemula*, 49 × 32)
c. *Dryopteris carthusiana* type (e.g. *D. carthusiana*, 56 × 35)
d. *Blechnum spicant* 63 × 38
e. *Woodsia* type (e.g. *W. ilvensis*, 43 × 35)
f. *Isoetes* (e.g. *I. lacustris*, 46 × 35)
g. *Asplenium* type (e.g. *A. septentrionale*, 45 × 33)
h. *Asplenium* type (e.g. *A. adiantum-nigrum*, 45 × 32)

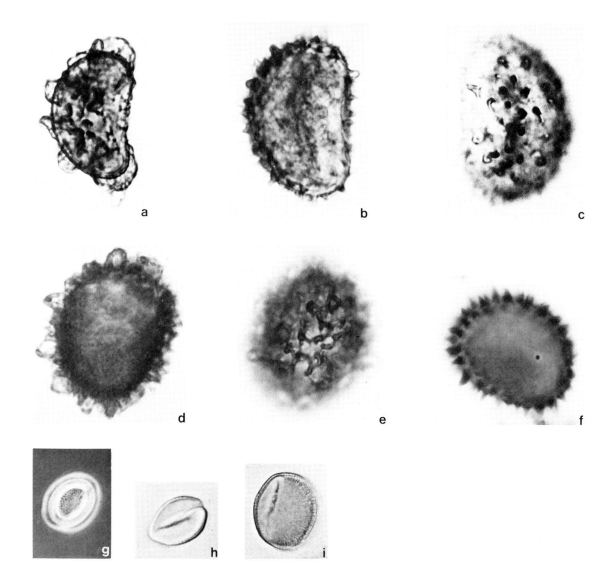

Plate 19
a. *Thelypteris dryopteris* 37 × 29
b. and c. *Thelypteris palustris* type (e.g. *T. palustris*, 48 × 33)
d. and e. *Polystichum type (e.g. P. lonchitis,* 40 × 29)
f. *Cystopteris* type (e.g. *C. fragilis*, 51 × 38)
g. (Ph) and h. *Tofieldia pusilla* 19 × 14
i. *Tamus communis* 32 × 25

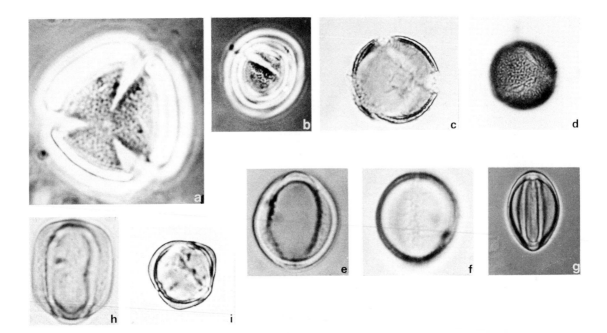

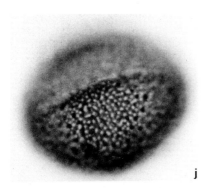

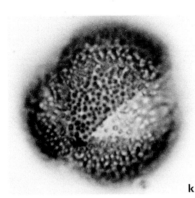

Plate 20
a. (Ph) and b. (Ph) *Melampyrum* (e.g. *M. pratense*, 22 × 21)
c. *Spergula* type (e.g. *S. arvensis*, 29 × 32)
d. *Spergula* type (e.g. *Spergularia media*, 25)
e. and f. *Trollius europaeus* 19 × 13
g. *Aconitum* (e.g. *A. napellus*, 27 × 16)
h. *Alchemilla* type (e.g. *Aphanes arvensis*, 22 × 18)
i. *Alchemilla* type (e.g. *Alchemilla glabra*, 22 × 24)
j. and k. *Glaucium* (e.g. *G. flavum*, 38 × 33)
l. and m. *Convolvulus arvensis* 48 × 51

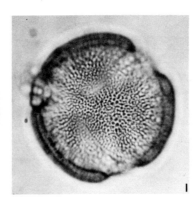

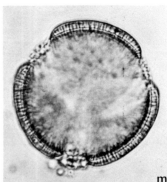

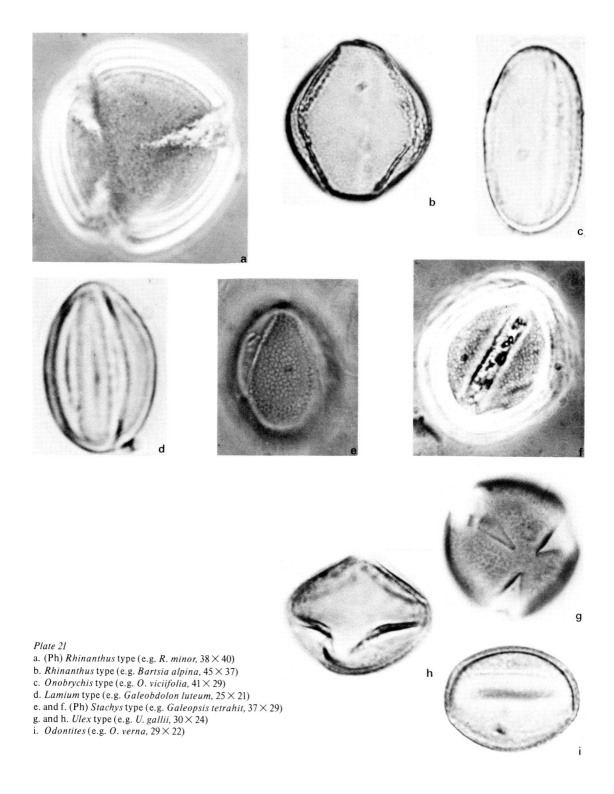

Plate 21
a. (Ph) *Rhinanthus* type (e.g. *R. minor*, 38 × 40)
b. *Rhinanthus* type (e.g. *Bartsia alpina*, 45 × 37)
c. *Onobrychis* type (e.g. *O. viciifolia*, 41 × 29)
d. *Lamium* type (e.g. *Galeobdolon luteum*, 25 × 21)
e. and f. (Ph) *Stachys* type (e.g. *Galeopsis tetrahit*, 37 × 29)
g. and h. *Ulex* type (e.g. *U. gallii*, 30 × 24)
i. *Odontites* (e.g. *O. verna*, 29 × 22)

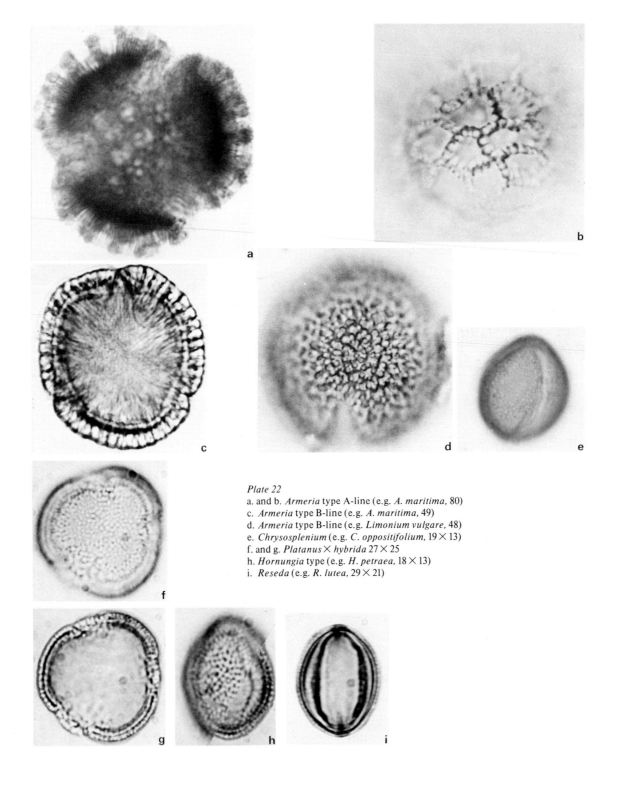

Plate 22
a. and b. *Armeria* type A-line (e.g. *A. maritima*, 80)
c. *Armeria* type B-line (e.g. *A. maritima*, 49)
d. *Armeria* type B-line (e.g. *Limonium vulgare*, 48)
e. *Chrysosplenium* (e.g. *C. oppositifolium*, 19 × 13)
f. and g. *Platanus × hybrida* 27 × 25
h. *Hornungia* type (e.g. *H. petraea*, 18 × 13)
i. *Reseda* (e.g. *R. lutea*, 29 × 21)

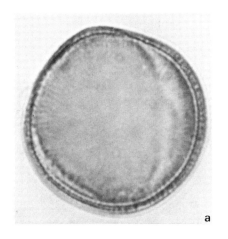

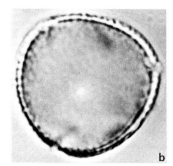

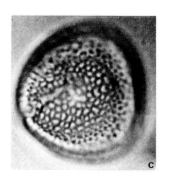

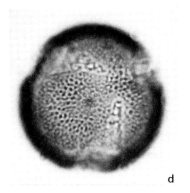

Plate 23
a. *Chelidonium majus* 27 × 25
b. and c. *Fraxinus excelsior* 38
d. *Helleborus* (e.g. *H. niger*, 38 × 27)
e. *Marrubium* (e.g. *M. vulgare*, 21)
f., g. and h. *Sinapis* type (e.g. *Sisymbrium officinale*, 19)
i., j. and k. *Salix* (e.g. *S. caprea*, 30 × 21)

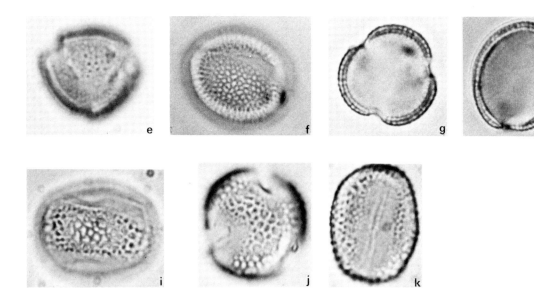

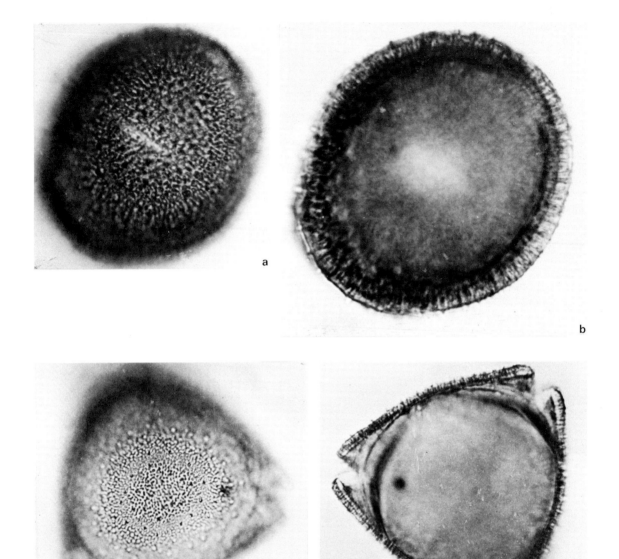

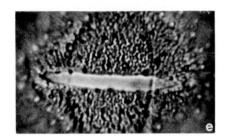

Plate 24
a. and b. *Succisa pratensis* 75
c., d. and e. *Lonicera* (e.g. *L. periclymenum*, 72)

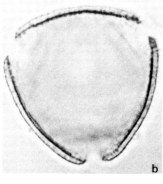

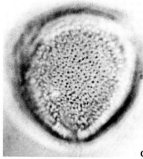

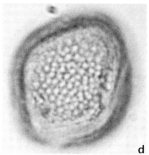

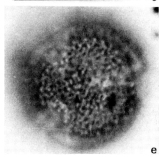

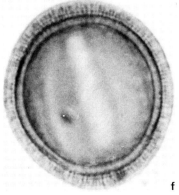

Plate 25
a. *Scabiosa columbaria* 54 × 45
b. and c. *Linnaea borealis* 43
d. and e. *Papaver* (e.g. *P. rhoeas*, 24 × 21)
f. and g. *Valerianella* (e.g. *V. locusta*, 33 × 30)
h. and i. *Valeriana* (e.g. *V. officinalis*, 48 × 41)

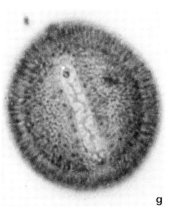

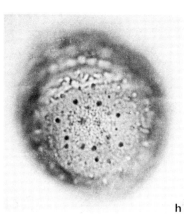

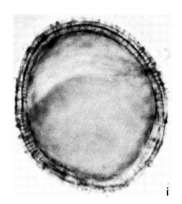

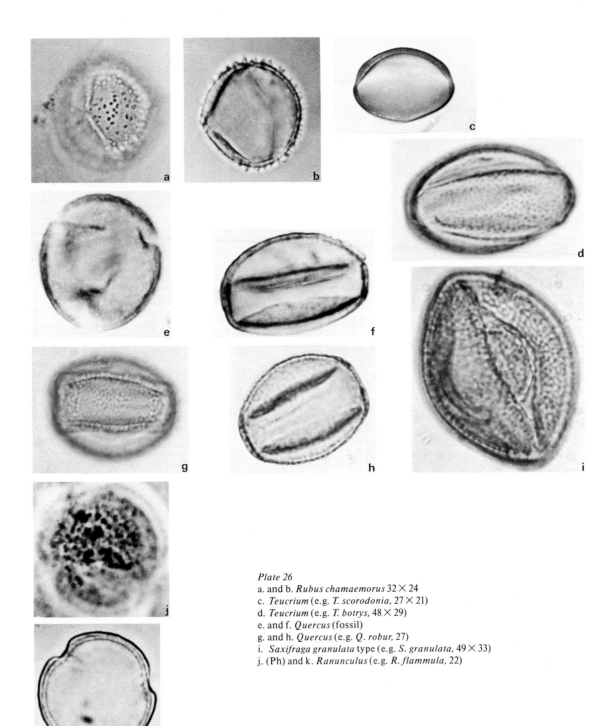

Plate 26
a. and b. *Rubus chamaemorus* 32 × 24
c. *Teucrium* (e.g. *T. scorodonia*, 27 × 21)
d. *Teucrium* (e.g. *T. botrys*, 48 × 29)
e. and f. *Quercus* (fossil)
g. and h. *Quercus* (e.g. *Q. robur*, 27)
i. *Saxifraga granulata* type (e.g. *S. granulata*, 49 × 33)
j. (Ph) and k. *Ranunculus* (e.g. *R. flammula*, 22)

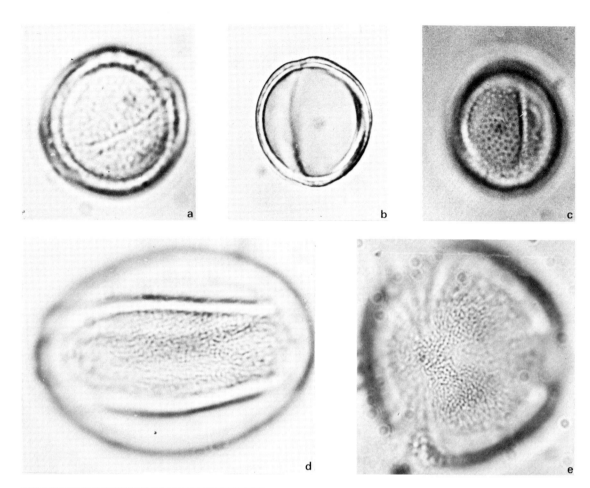

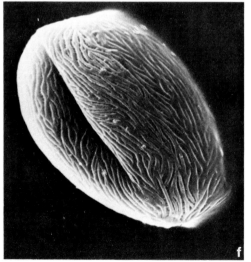

Plate 27
a. *Anemone* type (e.g. *A. appeninum*, 22)
b. and c. *Caltha* type (e.g. *C. palustris*, 24 × 25)
d., e. and f. (SEM) *Acer* (e.g. *A. pseudoplatanus*, 37 × 27)

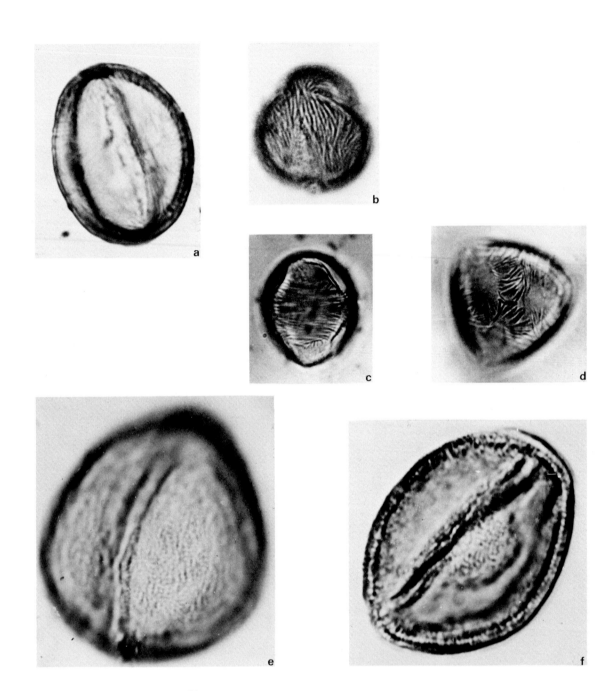

Plate 28
a. and b. *Saxifraga oppositifolia* type (e.g. *S. oppositifolia,*
    35 × 27)
c. and d. *Menyanthes trifoliata* 51 × 49
e. and f. *Dryas octopetala* 29 × 27

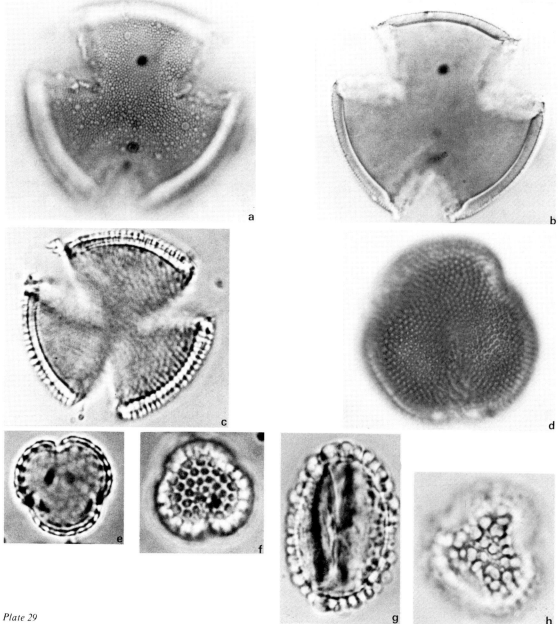

Plate 29
a. and b. *Linum bienne* type (e.g. *L. usitatissimum*, 56)
c. and d. *Linum catharticum* type (e.g. *L. catharticum*, 40)
e. and f. *Radiola linoides* 19 × 14
g. and h. *Ilex aquifolium* 32 × 25

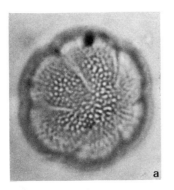

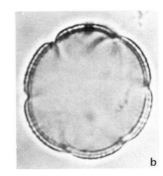

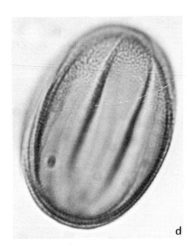

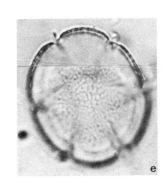

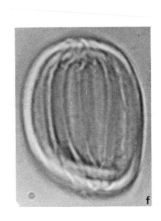

Plate 30

a. and b. *Mentha* type (e.g. *Lycopus europaeus*, 29)

c. *Impatiens* (e.g. *I. parviflora*, 35)

d. *Prunella* type (e.g. *Glechoma hederacea*, 35 × 22)

e. *Primula* (e.g. *P. vulgaris*, 24)

f. *Ephedra* (e.g. *E. distachya*, 30 × 24)

g. (SEM) and h. *Rubiaceae* (e.g. *Galium cruciata*, 16)

i. and j. *Hippuris vulgaris* 22

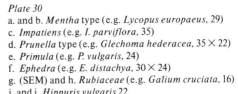

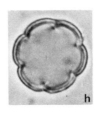

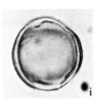

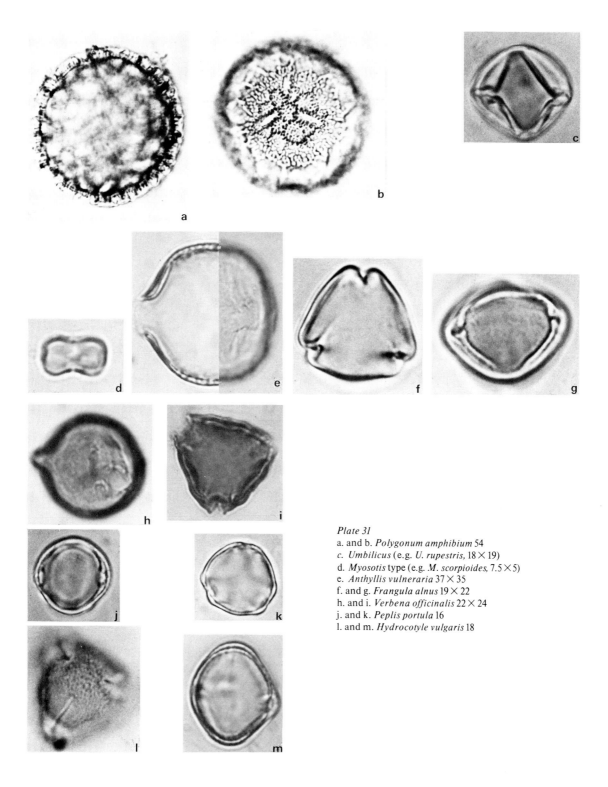

Plate 31
a. and b. *Polygonum amphibium* 54
c. *Umbilicus* (e.g. *U. rupestris*, 18 × 19)
d. *Myosotis* type (e.g. *M. scorpioides*, 7.5 × 5)
e. *Anthyllis vulneraria* 37 × 35
f. and g. *Frangula alnus* 19 × 22
h. and i. *Verbena officinalis* 22 × 24
j. and k. *Peplis portula* 16
l. and m. *Hydrocotyle vulgaris* 18

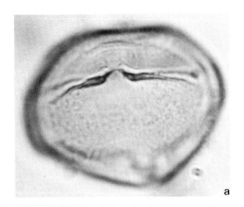

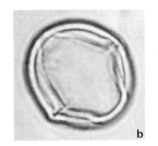

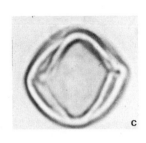

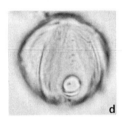

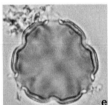

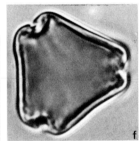

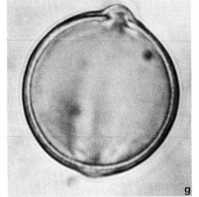

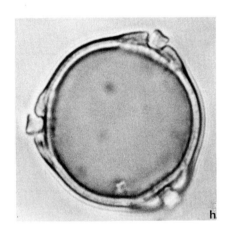

Plate 32
a. *Chamaepericlymenum suecicum* 33 × 29
b. *Solanum nigrum* 25 × 22
c. *Solanum dulcamara* 13 × 14
d. and e. *Lythrum* (e.g. *L. salicaria*, 30)
f. *Thelycrania sanguinea* 57 × 46
g. and h. *Poterium sanguisorba* 30
i. and j. *Filipendula* (e.g. *F. ulmaria*, 14)

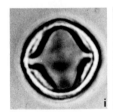

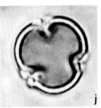

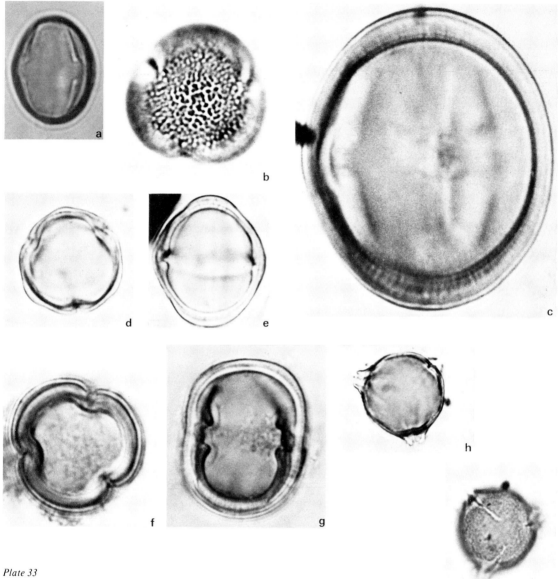

Plate 33
a. *Bupleurum* type (e.g. *B. falcatum*, 19 × 18)
b. *Polygonum bistorta* type (e.g. *P. viviparum*, 46 × 35)
c. *Polygonum bistorta* type (e.g. *P. bistorta*, 38 × 33)
d. and e. *Polygonum convolvulus* type (e.g. *P. convolvulus*,
    24 × 21)
f. and g. *Polygonum aviculare* 29 × 22
h. and i. *Hippophaë rhamnoides* 25

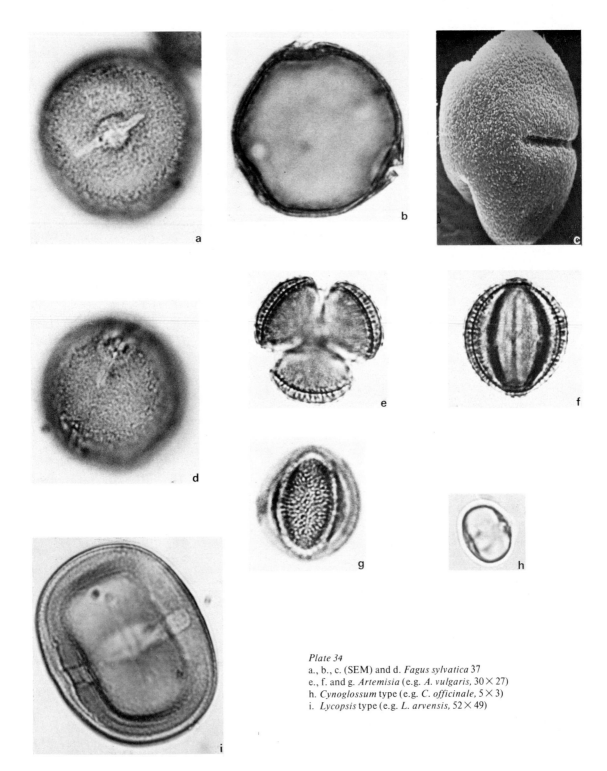

Plate 34
a., b., c. (SEM) and d. *Fagus sylvatica* 37
e., f. and g. *Artemisia* (e.g. *A. vulgaris*, 30 × 27)
h. *Cynoglossum* type (e.g. *C. officinale*, 5 × 3)
i. *Lycopsis* type (e.g. *L. arvensis*, 52 × 49)

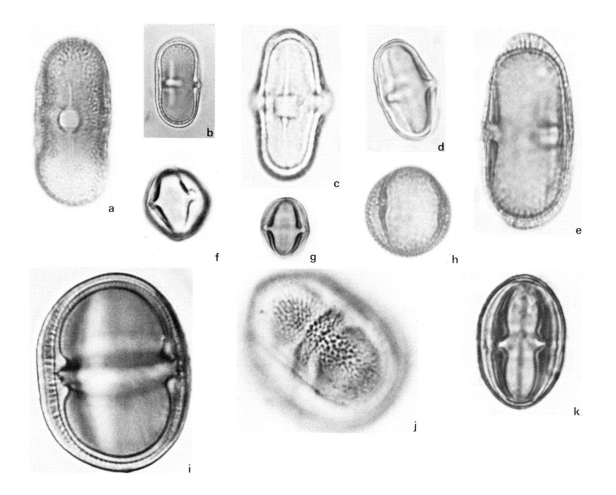

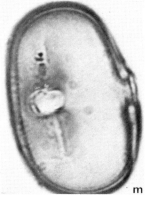

Plate 35
a. Umbelliferae 2 (e.g. *Myrrhis odorata*, 40 × 21)
b. Umbelliferae 2 (e.g. *Pastinaca sativa*, 32 × 20)
c. Umbelliferae 1 (e.g. *Smyrnium olusatrum*, 32 × 16)
d. Umbelliferae 3 (e.g. *Crithmum maritimum*, 22 × 13)
e. Umbelliferae 1 (e.g. *Heracleum sphondylium*, 35 × 18)
f. *Lotus* (e.g. *L. corniculatus*, 14 × 10)
g. *Castanea sativa* 18 × 13
h. *Oxyria* type (e.g. *O. digyna*, 25)
i. and j. *Centaurea cyanus* 35 × 25
k. *Glaux maritima* 25 × 18
l. and m. *Vicia cracca* type (e.g. *V. cracca*, 49 × 33)

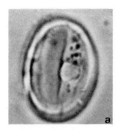
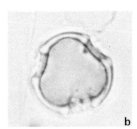
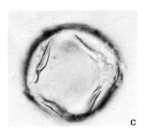
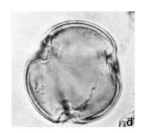

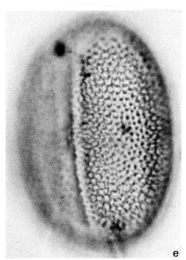
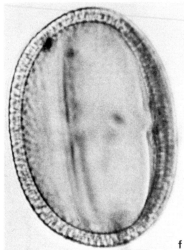
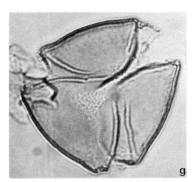

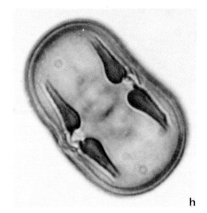

Plate 36
a. and b. *Aesculus hippocastanum* 29 × 18
c. and d. *Crataegus* type (e.g. *C. monogyna*, 32 × 27)
e. and f. *Fagopyrum esculentum* 49 × 41
g. *Viola palustris* type (e.g. *V. palustris*, 33 × 26)
h. *Lathyrus* type (e.g. *L. pratensis*, 41 × 27)

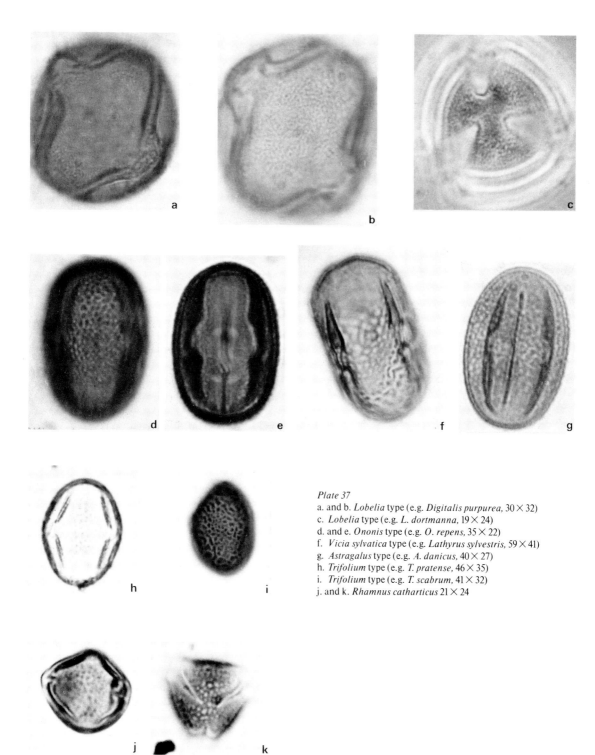

Plate 37
a. and b. *Lobelia* type (e.g. *Digitalis purpurea*, 30 × 32)
c. *Lobelia* type (e.g. *L. dortmanna*, 19 × 24)
d. and e. *Ononis* type (e.g. *O. repens*, 35 × 22)
f. *Vicia sylvatica* type (e.g. *Lathyrus sylvestris*, 59 × 41)
g. *Astragalus* type (e.g. *A. danicus*, 40 × 27)
h. *Trifolium* type (e.g. *T. pratense*, 46 × 35)
i. *Trifolium* type (e.g. *T. scabrum*, 41 × 32)
j. and k. *Rhamnus catharticus* 21 × 24

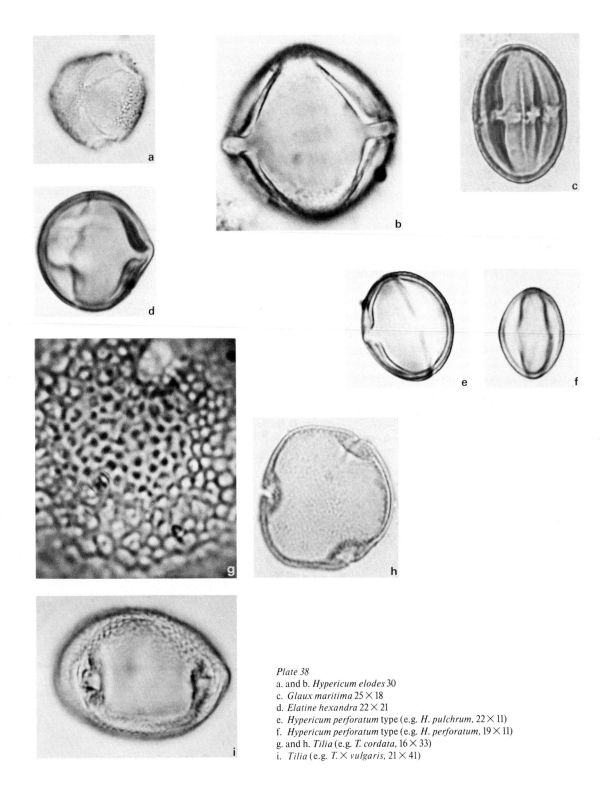

Plate 38
a. and b. *Hypericum elodes* 30
c. *Glaux maritima* 25 × 18
d. *Elatine hexandra* 22 × 21
e. *Hypericum perforatum* type (e.g. *H. pulchrum*, 22 × 11)
f. *Hypericum perforatum* type (e.g. *H. perforatum*, 19 × 11)
g. and h. *Tilia* (e.g. *T. cordata*, 16 × 33)
i. *Tilia* (e.g. *T. × vulgaris*, 21 × 41)

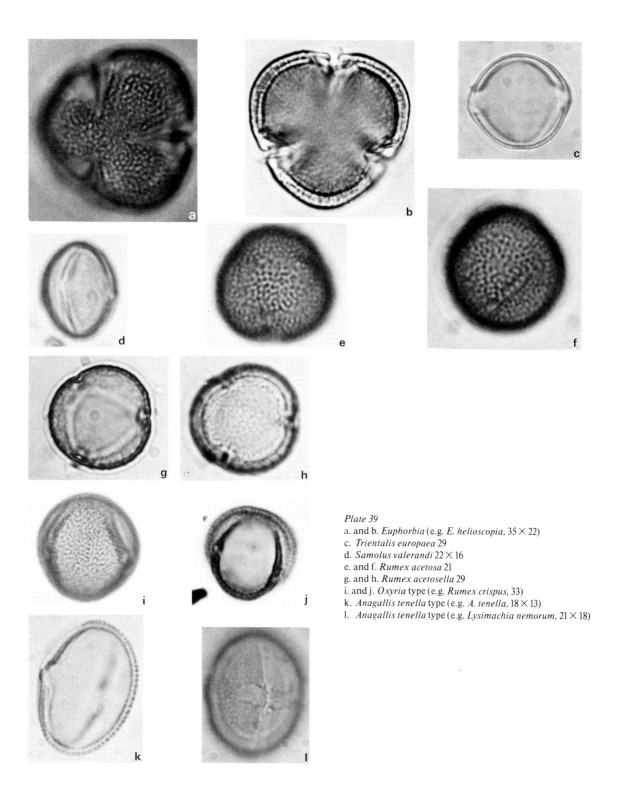

Plate 39
a. and b. *Euphorbia* (e.g. *E. helioscopia*, 35 × 22)
c. *Trientalis europaea* 29
d. *Samolus valerandi* 22 × 16
e. and f. *Rumex acetosa* 21
g. and h. *Rumex acetosella* 29
i. and j. *Oxyria* type (e.g. *Rumex crispus*, 33)
k. *Anagallis tenella* type (e.g. *A. tenella*, 18 × 13)
l. *Anagallis tenella* type (e.g. *Lysimachia nemorum*, 21 × 18)

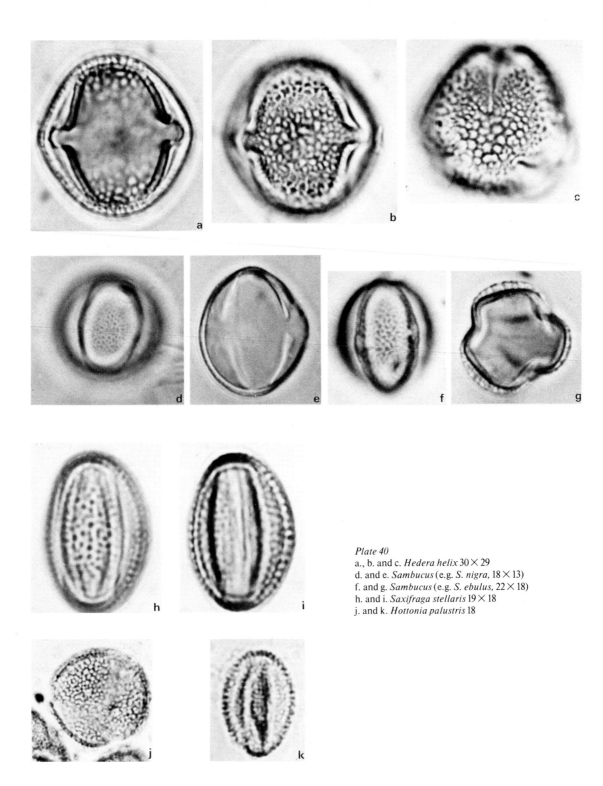

Plate 40
a., b. and c. *Hedera helix* 30 × 29
d. and e. *Sambucus* (e.g. *S. nigra,* 18 × 13)
f. and g. *Sambucus* (e.g. *S. ebulus,* 22 × 18)
h. and i. *Saxifraga stellaris* 19 × 18
j. and k. *Hottonia palustris* 18

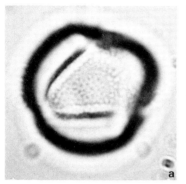

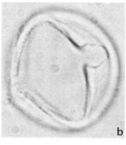

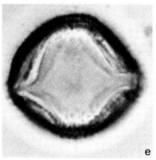

Plate 41
a. and b. *Linaria* (e.g. *L. vulgaris*, 19)
c. and d. *Gentianella* (e.g. *G. amarella*, 41 × 32)
e. and f. *Viburnum* (e.g. *V. lantana*, 16)
g. and h. *Lysimachia vulgaris* type
 (e.g. *L. vulgaris*, 25 × 21)

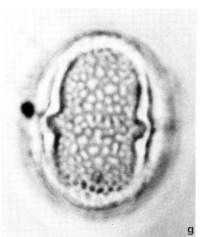

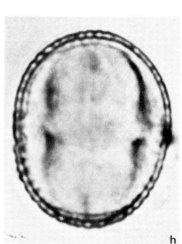

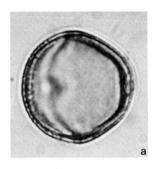

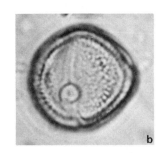

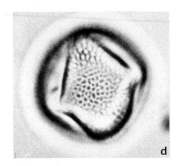

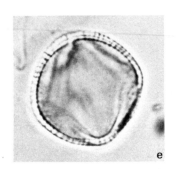

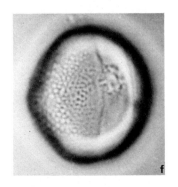

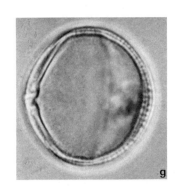

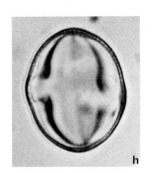

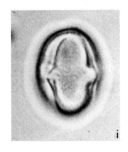

Plate 42
a. and b. *Parnassia palustris* 24
c. *Euonymus europaeus* 24
d. and e. *Scrophularia* type
   (e.g. *Verbascum thapsus*, 40 × 38)
f. and g. *Gentiana verna* 40 × 32
h. and i. *Anagallis arvensis* 25 × 22

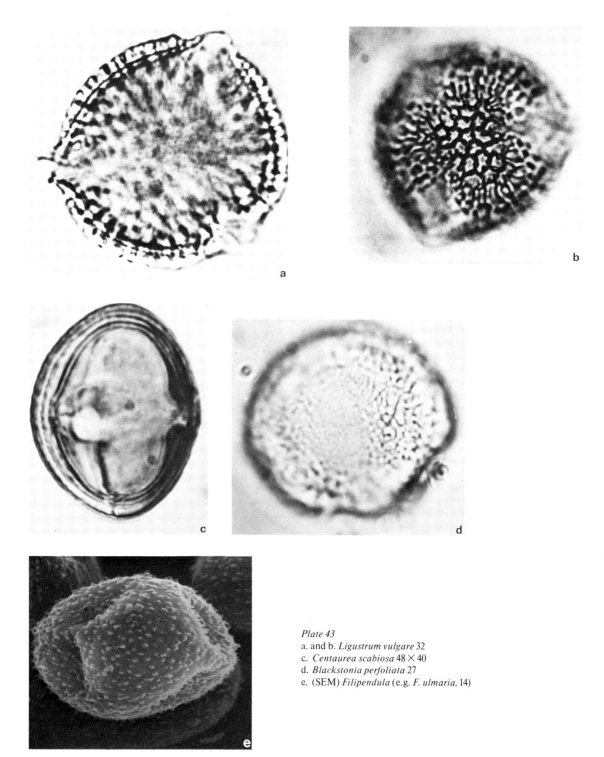

Plate 43
a. and b. *Ligustrum vulgare* 32
c. *Centaurea scabiosa* 48 × 40
d. *Blackstonia perfoliata* 27
e. (SEM) *Filipendula* (e.g. *F. ulmaria*, 14)

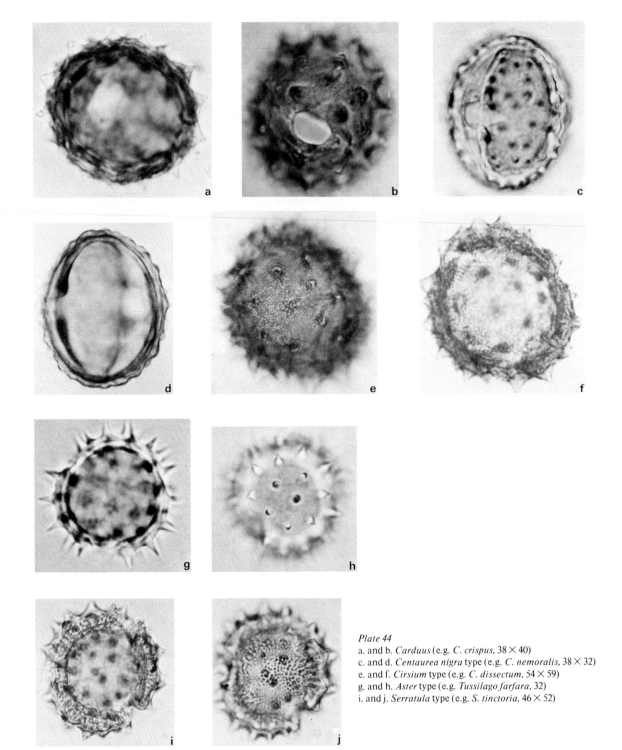

Plate 44
a. and b. *Carduus* (e.g. *C. crispus*, 38 × 40)
c. and d. *Centaurea nigra* type (e.g. *C. nemoralis*, 38 × 32)
e. and f. *Cirsium* type (e.g. *C. dissectum*, 54 × 59)
g. and h. *Aster* type (e.g. *Tussilago farfara*, 32)
i. and j. *Serratula* type (e.g. *S. tinctoria*, 46 × 52)

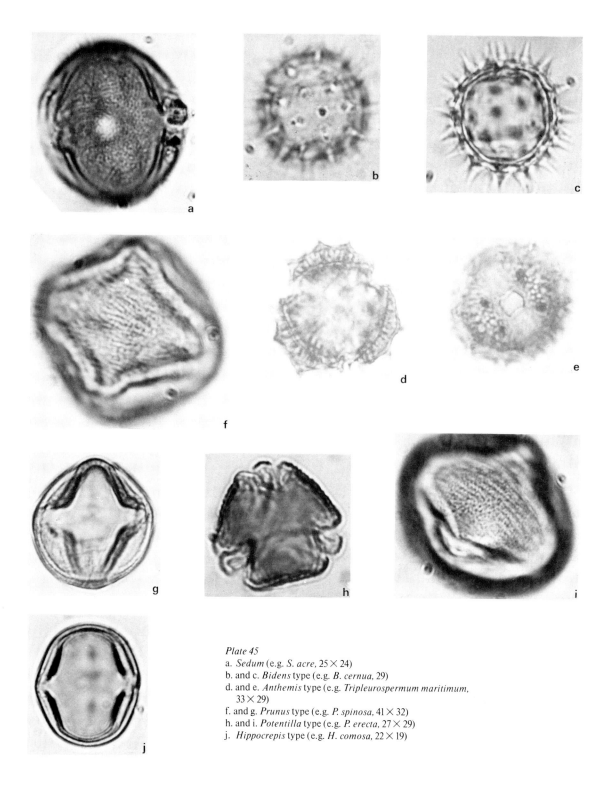

Plate 45
a. *Sedum* (e.g. *S. acre*, 25 × 24)
b. and c. *Bidens* type (e.g. *B. cernua*, 29)
d. and e. *Anthemis* type (e.g. *Tripleurospermum maritimum*,
    33 × 29)
f. and g. *Prunus* type (e.g. *P. spinosa*, 41 × 32)
h. and i. *Potentilla* type (e.g. *P. erecta*, 27 × 29)
j. *Hippocrepis* type (e.g. *H. comosa*, 22 × 19)

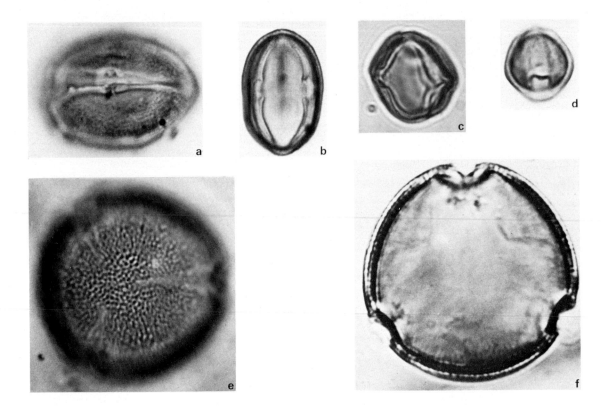

Plate 46
a. and b. *Gentiana pneumonanthe* 24×21
c. and d. *Sedum* (e.g. *S. roseum*, 16)
e. and f. *Helianthemum* (e.g. *H. chamaecistus*, 40×38)
g. and h. *Helianthemum* (e.g. *H. oelandicum*, 40×32)

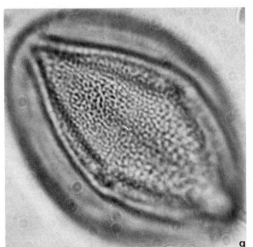

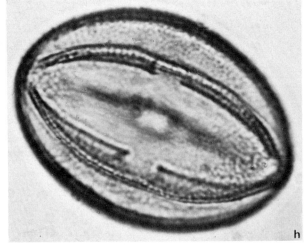

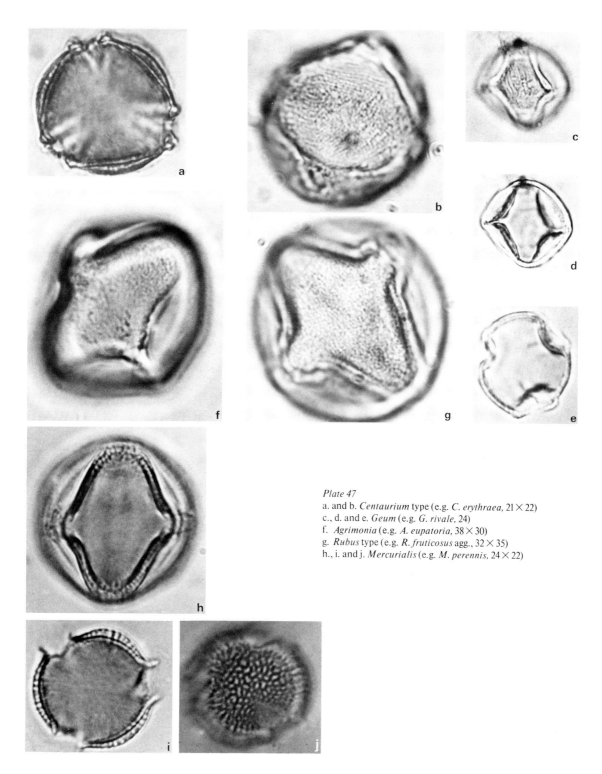

Plate 47
a. and b. *Centaurium* type (e.g. *C. erythraea*, 21 × 22)
c., d. and e. *Geum* (e.g. *G. rivale*, 24)
f. *Agrimonia* (e.g. *A. eupatoria*, 38 × 30)
g. *Rubus* type (e.g. *R. fruticosus* agg., 32 × 35)
h., i. and j. *Mercurialis* (e.g. *M. perennis*, 24 × 22)

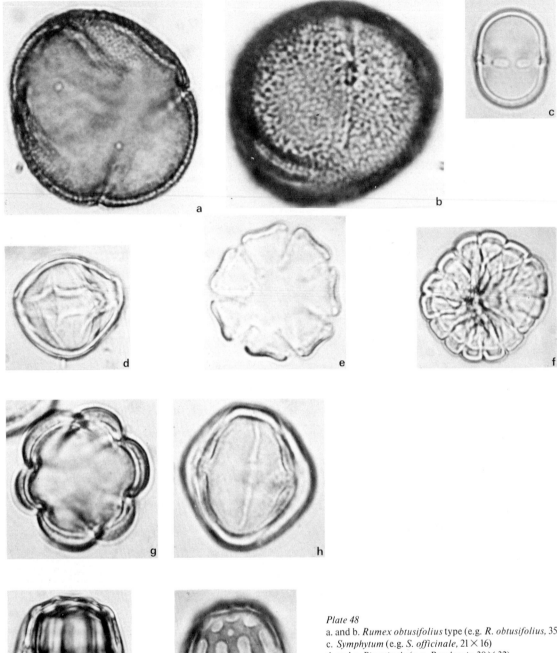

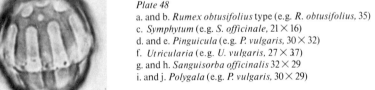

Plate 48
a. and b. *Rumex obtusifolius* type (e.g. *R. obtusifolius*, 35)
c. *Symphytum* (e.g. *S. officinale*, 21 × 16)
d. and e. *Pinguicula* (e.g. *P. vulgaris*, 30 × 32)
f. *Utricularia* (e.g. *U. vulgaris*, 27 × 37)
g. and h. *Sanguisorba officinalis* 32 × 29
i. and j. *Polygala* (e.g. *P. vulgaris*, 30 × 29)

2 b No differentiated area around each porus. Edge of porus either well defined or ragged and diffuse: 7

3 a Grain with > 40 pori. Exine tectate, tectum without perforations: Chenopodiaceae

3 b Grain with < 40 pori. Exine tectate-perforate or stratification indistinct: 4

4 a Pori not protruding in an optical section of the grain, exine thinner at the porus edge than in the mesoporium. At least some of the columellae coarse, tectum perforate with minute echinae between the perforations: Caryophyllaceae (plus *Salsola*)

[For differentiation of types within this family see Faegri and Iversen, 1964; Chanda, 1962.]

4 b Margins of pori slightly protruding in an optical section of the grain. Exine ± thicker at the porus edge but not markedly thinner than in the middle of the mesoporium. Columellae uniform and fine: 5

5 a Differentiated area around each porus is a result of dissolution or fragmentation of the nexine. Exine covered by very regularly spaced microechinae, grain with 6–20 pori, sometimes aggregated towards one part of the grain: *Juglans*

5 b Differentiated area around each porus is the result of thickening of the sexine (i.e. an annulus). This annulus is visible as a solid ring to each porus. Pori ± operculate, ± microechinate: 6

6 a Grain with 6–11 pori (the number is commonly more than 8), distinctly microechinate (phase): *Plantago lanceolata*

6 b Grain with < 8 pori, microechinae indistinct (phase): *Plantago coronopus*

[See Faegri and Iversen, 1974 and also page 54 of this volume.]

7 a Exine thickest in the middle of the mesoporium and thinnest near the pori: 8

7 b Exine of uniform thickness all over the grain surface. Sculpturing consists of regularly spaced microechinae: 9

8 a Grain large (> 40 μm). Exine 5–7 μm thick, tectate-perforate (perforations minute). Tectum supported by very coarse columellae which are branched at two-thirds of their length: *Calystegia* type

8 b Grains < 40 μm. Exine 2 μm or less, a large area around each porus where the columellae are shorter, finer and more regularly distributed than those columellae in the middle of the mesoporium: *Alisma* type

[Includes *Alisma*, *Baldellia* and *Luronium*.]

9 a Pori situated in slightly sunken areas of the exine, giving a dimpled appearance to the grain. Porus margin very diffuse, porus membrane granulate: *Thalictrum* type

[Includes *T. flavum*, *T. minus* and *T. alpinum*.]

9 b Pori not situated in sunken areas, but are flat on the surface of the grain. Pori circular, faint, but margin clearly demarcated, sometimes with isolated large granules in the centre of the aperture: *Plantago maritima*

[See also note on page 54.]

10 a Grains echinate or verrucate: 11

10 b Grains reticulate or rugulate-striate: 19

11 a Grain echinate or microechinate: 12

11 b Grain verrucate (with low, broad-based lumps, black in phase contrast): 15

12 a Grain large (~ 100 μm) with long (> 6 μm) sharply pointed echinae: *Malva* type

[Includes *Malva*, *Lavatera* and *Althaea*.]

12 b Grain < 60 μm, echinae < 6 μm: 13

13 a Pori well defined and generally each with an annulus (sexine thinner and without columellate structure near the pori). Echinae 0·5 μm or less, grains tectate-perforate: Caryophyllaceae

[For differentiation of types within this family Faegri and Iversen, 1964; Chanda, 1962.]
Faegri and Iversen, 1964; Chanda, 1962.]

13 b No well-defined circular edge to porus and no annulus present. Echinae slightly more prominent (0·5 μm–> 1 μm): 14

14 a Pori operculate, operculum covered with echinae in the same way as the rest of the exine. Echinae broad based and conical: *Sagittaria*

[See also *Ranunculus arvensis*, page 56.]

14 b Pori not operculate, echinae less broad based and conical: *Koenigia*

15 a Each porus with an area of thickened sexine around it (annulus). Grains ± operculate: 16

15 b No such annulus to each porus. Grains never operculate, but with granulate porus-membrane: 17

16 a Grain with 6–11 pori (although the number is commonly more than 8), microechinae visible—especially between the verrucae (use phase contrast): *Plantago lanceolata*

16 b Grain with < 8 pori, microechinae indistinct (use phase contrast): *Plantago coronopus*

[See also page 54 and Faegri and Iversen, 1964, check with type slides.]

17 a Grain with ⩾ 8 pori, commonly > 30 μm in size. Verrucae appear small in comparison to grain size (see photograph): *Littorella*

17 b Grains with ⩽ 8 pori, may or may not be < 30 μm in diameter. Verrucae appear larger in relation to grain size (see Plate 11): 18

18 a Pori sharply delimited. Sculpturing consists of

low verrucae with distinct and regularly spaced microechinae. Columellae visible in phase contrast as a dense carpet under the verrucae and microechinae: *Plantago maritima*

18 b Pori not so sharply delimited (presence is shown by the cessation of verrucate sculpturing). Columellae fine to invisible. Microechinae ± visible: *Plantago media/major*

[See also page 54.]

19 a Grains with rugulae or winding striae, muri wide ($\sim 1.5 \mu$m). Columellae sit in a reticulate pattern underneath the striations (see the results of an 'LO' analysis). Pori > 50 in number: *Polemonium*

19 b Grains eureticulate. Pori < 50 in number: 20

20 a Each porus small, so that it just fills the bottom of a lumen. Reticulum thus not broken up by the pori: 21

20 b Pori not in the lumina of the reticulum. Porus diameter much greater than that of the lumina, pori thus interrupt the reticulum pattern: 22

21 a Non-porate lumina with baculate or granulate floors: *Polygonum persicaria* type.
[Includes *P. persicaria, P. lapathifolium, P. mite* and *P. hydropiper*.]

21 b Non-porate lumina without bacula or granules: *Daphne*

22 a Well-defined circular edge to porus. Annulus present (here a thinning and consolidation of the sexine): *Caryophyllaceae*

22 b No well-defined circular edge to porus. Annulus not present. Porus demarcated by the cessation of the reticulate sculpturing: *Buxus sempervirens*

N.B. Irregularly polypantoporate grains are found in some species of genera which in Britain are otherwise trizonocolpate or trizonocolporate, e.g. *Linum anglicum, Rorippa sylvestris, Papaver argemone, Ranunculus arvensis* (see Plate 13). *Trientalis europaea* produces polypantoporate and trizonocolporate to polyzonocolporate grains.

Syncolpate (Plate 15)

1 a Colpi running laterally in wide spirals around the grain. Exine with scattered small echinae: *Eriocaulon*

1 b Colpi running meridionally and fused in the polar regions of the grain. Sculpture never echinate: 2

2 a With two colpi which are fused at the poles to form a complete ring around the grain. Colpi with very ragged edges. Grains sometimes split into two halves: *Pedicularis*

2 b With three or more colpi, each colpus in the polar region bifurcated and fused with neighbouring two colpi: 3

3 a Grain > 30 μm, sculpturing coarsely rugulate: *Nymphoides*

3 b Grain < 30 μm, sculpturing minutely eureticulate: *Primula farinosa*

N.B. Some species of *Ephedra* are syncolpate at the poles (see page 63).

Monocolpate (Plates 16 to 19)

1 a Grains echinate or reticulate, or tectate-perforate or faintly verrucate-rugulate. Grain wall with or without columellate structure: 2

1 b Grains bean-shaped; smooth or possessing two coats, the outer variously folded, wrinkled or spiny. Grain wall without columellate structure: 8

2 a With long (5–8 μm) echinae scattered over the surface. Colpus may be indistinct because it is operculate: *Nuphar*

[See page 50.]

2 b Grains psilate to reticulate but never echinate, colpus non-operculate: 3

3 a Grain semitectate, eureticulate: 4

3 b Grains tectate, perforate or with fine, indistinct sculpturing: 7

4 a Grain < 40 μm in length: 5

4 b Grain > 40 μm in length: 6

5 a Lumina becoming markedly smaller in size towards the colpus (colpus thus has a definite margo). Columellae easily visible under × 1000: *Butomus*

5 b Lumina not becoming markedly smaller towards the colpus. Columellae invisible (× 1000) or only just visible under phase: *Narthecium*

6 a With adjacent lumina of widely differing sizes. Muri often with more than one row of columellae supporting them (× 1000 and phase contrast). Lumina becoming smaller towards colpus: *Scilla* type

[Includes *Scilla* and *Lloydia*.]

6 b With adjacent lumina of roughly the same size. Muri with only one row of coarse columellae supporting each one: *Iris*

7 a Grain tectate with perforations: *Convallaria* type
[Includes *Convallaria, Maianthemum, Polygonatum* and *Paris*.]

7 b Grain psilate or faintly verrucate-rugulate on top of the tectum (× 1000, phase contrast). Columellae visible as dense carpet: *Allium* type

[Includes *Allium*, *Ruscus* and *Polygonatum*.]

8 a Grain completely smooth, bean-shaped: Polypodiaceae

8 b Grain never completely smooth. Bean shape often obscured by outer coat which is variously wrinkled, echinate or verrucate: 9

9 a Outer coat (perine) closely attached to inner coat, not thrown up into distinct *hollow* elevations such as echinae, ridges or sacci: 10

9 b Perine separated from inner coat either as a whole, or by hollow elevations: 11

10 a With large, undulating, solid verrucae. Individual verrucae > 3 μm high. Spore large, > 60 μm in length: *Polypodium vulgare (sensu-lato)*

10 b With small gemmae verrucae or scabrae. Spore < 60 μm in length: *Athyrium filix-femina*

11 a Perine either entirely detached from the inner coat, or thrown into sacs and wrinkles and attached at a few points: 12

11 b Perine closely attached to the inner coat except where it is thrown up into spines or ridges: 18

12 a Perine either unwrinkled or thrown into a few loose wrinkles (usually up to four on one side): 13

12 b Perine thrown into many sacci or folds (usually more than four on one side): 15

13 a Perine densely covered in small echinae (1·5–2 μm long): *Dryopteris dilatata*

13 b Perine smooth or granulate but not densely covered in echinae: 14

14 a Perine bulging like a sac at the ends and around the sides of the colpus. Perine rarely wrinkled: *Isoetes*

14 b Perine not bulging in the region of the colpus, commonly with a few wrinkles: *Blechnum spicant*

15 a Perine surface smooth or with only very faint granulations: 16

15 b Perine surface echinate, granulate or foveolate: 17

16 a With straight ridges or folds, which tend to intersect or anastomose to form a very coarse reticulum: *Woodsia* type

[Includes *Woodsia ilvensis*, *Woodsia alpina* and *Athyrium alpestre*.]

16 b With curving and twisting ridges or bulging sacci, rarely intersecting and never anastomosing: *Dryopteris filix-mas* type

[Includes *D. filix-mas*, *D. abbreviata*, *D. borreri*, *D.*

aemula, *Thelypteris robertianum*, *Thelypteris limbosperma*, *Cystopteris fragilis* ssp. *dickeana*.]

17 a Perine surface covered in small echinae: *Dryopteris carthusiana* type

[Includes *D. carthusiana* and *D. cristata*.]

17 b Perine surface granulate to foveolate, but not echinate: *Thelypteris dryopteris*, *Thelypteris phegopteris* (Compare with type slides to separate these two.)

18 a Perine thrown into coarse, anastomosing ridges. Small echinae are present on top of the ridges: *Asplenium* type

[Includes *A. adiantum-nigrum*, *A. viride*, *A. trichomanes*, *A. ruta-muraria*, *A. septentrionale*, *Ceterach officinarum*.]

18 b Perine thrown into echinae, pointed processes or papillae rather than ridges: 19

19 a With dense, irregular-pointed processes— their bases joined by low ridges: *Polystichum* type

[Includes *P setiferum*, *P. lonchitis*, *Asplenium marinum* and *Phyllitis scolopendrium*.]

19 b Discrete echinae present, faint lines sometimes joining their bases: 20

20 a Echinae broad based and narrowing gradually above the base—thus appearing parallel-sided to triangular: *Cystopteris fragilis* type

[Includes *C. fragilis* and *C. montana*.]

20 b Echinae narrowing quickly above their bases, faint raised lines may run between the bases of the echinae: *Thelypteris palustris* type

[Includes *T. palustris*, *Asplenium marinum* and *Phyllitis scolopendrium*. See also *Asplenium* type, page 57 and *Polystichum* type, page 57.]

For discussions and descriptions of fern spore morphology see Sorsa (1964) and Knox (1951).

Dizonocolpate (Plate 19)

Very few pollen grains have two colpi as their normal compliment but the two-colpate condition does sometimes occur in grains which are normally three- (or more) colpate.

1 a Reticulate, lumina becoming markedly smaller towards the colpi. Columellae visible (× 1000 phase contrast), coarsest and longest in area between colpi: *Tamus*

1 b Reticulate, lumina near colpi only slightly smaller than those away from colpi, columellae invisible. Exine thin (1 μm thick). Grain < 25 μm: *Tofieldia*

Trizonocolpate

This section is subdivided on the basis of sculpture type (see page 41)

PSILATE, PERFORATE (page 58)

RETICULATE (page 59)

ECHINATE (page 60)

SCABRATE-VERRUCATE (page 60)

RUGULATE-STRIATE (page 61)

BACULATE (p. 61)

CLAVATE (page 62)

Trizonocolpate: Psilate or Perforate (Plates 20 and 21)

1 a   Grain with exine > 2·5 μm, with fairly coarse columellae (1 μm or more thick): 2

1 b   Grain with exine < 2·5 μm, columellae fine to indistinguishable (much less than 1 μm thick): 5

2 a   Grain with exine thickest and columellae longest in the mesocolpium. Grain triangular-obtuse in polar view, rectangular-obtuse in equatorial view: *Alchemilla* type

    [Includes *Alchemilla* and *Aphanes*; most *Alchemilla* species produce deformed grains, *Aphanes arvensis* (Plate 18b) produces well-formed grains.]

2 b   Grain with exine of uniform thickness all over the mesocolpium: 3

3 a   Columellae branched at two-thirds of their length. Grain > 40 μm in length. Perforations in the tectum minute and closely spaced: *Convolvulus*

    [See also page 66. *Polygonum bistorta* type may key out here if pori are indistinct.]

3 b   Columellae unbranched, grains < 40 μm. Perforations not so minute and closely spaced: 4

4 a   Grain circular in polar and equatorial view. Colpi narrow, often nearly meeting at the poles so that apocolpium is very small. Columellae never arranged in a reticulum in the mesocolpium: *Spergula* type

    [Includes *Spergula* and *Spergularia*. See also page 61.]

4 b   Grain often elliptic or circular in equatorial view. Colpi wide and granulate. Apocolpium larger. Columellae may be arranged in a reticulum in the mesocolpium (infrareticulate): *Glaucium*

    [See also page 60.]

5 a   Columellae indistinct at × 1000, even in phase contrast. Grains < 25 μm: 6

5 b   Columellae distinct black dots at × 1000,

especially in phase contrast. Grains may or may not be > 25 μm: 9

6 a   Grains minutely micro-reticulate at × 1000 and in phase contrast: 7

6 b   Grains not microreticulate, critical observation at × 1000 and phase contrast shows grain either striate or completely smooth: 8

7 a   Grain > 18 μm: *Saxifraga nivalis*

7 b   Grain < 18 μm and in many cases < 12 μm: *Chrysosplenium*

8 a   With very fine striae running across the mesocolpium in an equatorial direction. Colpus edge very ragged in phase contrast: *Trollius*

8 b   Grain completely psilate, exine slightly thicker at the poles than elsewhere on the grain: *Aconitum*

9 a   Exine thickest, columellae longer and more spaced out in the region of the mesocolpium near to the colpus, and in the apocolpium. Columellae short and densely packed in the middle of the mesocolpium and on the edge of the colpus (see Plate 18 *l* and *m*). Characteristic shape in polar view, optical section: *Melampyrum*

9 b   Columellae of equal length and even distribution all over the grain (i.e. not more spaced out near colpus): 10

10 a   Colpi wide, with granulate membrane, mesocolpium bulging, grain < 23 μm: *Frankenia*

    [See also *Trollius*, p. 58.]

10 b   Colpi narrow, often with torn and ragged edges, mesocolpium not so curved, grains may or may not be > 23 μm: 11

11 a   Grains large (> 30 μm), circular in both polar and equatorial views. Colpi thin and crack-like with ragged edges. Columellae quite fine, forming a dense carpet all over the grain: *Rhinanthus* type

    [Includes *Rhinanthus*, *Bartsia*, *Euphrasia* and *Veronica*. Compare *Viola palustris* type, page 66.]

11 b   Grains < 30 μm, elliptic in equatorial view. Colpi ± ragged-edged. Grain may appear minutely microreticulate in phase, often with pattern much coarser at the poles: *Lamium* type

    [Includes *Lamium*, *Galeobdolon*, *Stachys* and *Scutellaria*.]

                        Odontites type, *Spergula* type

                          [See also page 61.]

N.B. These grains have not been separated as we feel that reference to type slides is the only way in which this may be achieved.

Trizonocolpate: Reticulate (Plates 21 to 23)

1 a   Grains suprareticulate, i.e. the muri of the reticulum are walls which sit on top of a tectum. The reticulum is thus independent of the pattern of the columellae (see Fig. 4.8). Columellae are not always visible: 2

1 b   Grains eureticulate, i.e. the muri of the reticulum are walls which join the heads of the columellae. Distribution of columellae commonly dependent on the pattern of the reticulum (see Fig. 4.9). Columellae visible: 5

2 a   Exine > 3 μm thick, grain > 40 μm, columellae coarse and branched at two-thirds their length. Numerous tiny perforations of the tectum are visible in the lumina of the suprareticulum: *Convolvulus*

2 b   Exine < 3 μm thick, grain < 40 μm, columellae fine and unbranched: 3

3 a   Grains as long as broad, in equatorial view circular to rhombic-obtuse: *Ulex* type

     [Includes *Ulex, Genista, Cytisus* and *Sarothamnus.*]

3 b   Grains longer than broad, in equatorial view ± rhombic but never circular: 4

4 a   Each colpus narrow and crack-like, with granules in one line in the middle of it. Reticulum very regular with nearly isodiametric lumina, columellae fine to invisible. In equatorial view grain shape elliptic to rectangular-obtuse: *Onobrychis* type

     [Includes *Onobrychis, Hedysarum.*]

4 b   Each colpus boat-shaped, i.e. widest at the equator and narrowing to a point at each pole, colpus membrane bearing scattered granules. Reticulum lumina ± isodiametric. In equatorial view grain shape elliptic to rhombic-obtuse: *Stachys* type

     [Includes *Stachys, Galeopsis, Lamium, Ajuga* and *Scutellaria.*]

5 a   Grains microreticulate, i.e. with lumina ≤ 1 μm in width: 6

5 b   Grains not microreticulate, i.e. with some lumina > 1 μm in width: 10

6 a   Colpi wide, with granulate covering membrane: 7

6 b   Colpi narrow, thus no colpus membrane visible: 8

7 a   Colpi long and boat-shaped, i.e. widest at equator and narrowing to pointed ends at the poles. Edge of colpus well defined, neat. Grain elliptic to rhombic-obtuse in equatorial view: *Reseda*

7 b   Colpi short, parallel sided with diffuse rounded ends. Edge of colpus ill-defined, ragged. Grain elliptic to circular in equatorial view: *Platanus*

8 a   Columellae distinct, longest in the centre of the mesocolpium (i.e. slightly shorter at poles and at colpi margins): *Hornungia* type

     [Includes *Hornungia, Capsella* and others. See Erdtman, Praglowski and Nilsson, 1963.]

8 b   Columellae indistinct (invisible without acetolysis), same length all over the grain surface. Apocolpium very small: 9

9 a   Grain < 18 μm in length. Apocolpium with lumina so small as to appear tectate-perforate: *Chrysosplenium*

9 b   Grain > 18 μm in length. Apocolpium reticulate: *Saxifraga nivalis*

10 a   Exine > 4 μm thick, grains > 40 μm in size: 11

10 b   Exine < 4 μm thick, grains < 40 μm in size: 13

11 a   Reticulum very coarse, lumina in the mesocolpium ≥ 7 μm in width, row of echinae on the top of each murus. Exine ≥ 10 μm in thickness: *Armeria* type, A-line.

     [Includes *Armeria* and *Limonium*. The pollen of *Plumbaginaceae* is dimorphic. 'A-line' pollen is derived from flowers which possess smooth styles, while 'B-line' pollen is derived from flowers with papillose styles.]

11 b   Reticulum less coarse, i.e. lumina < 7 μm in width. Exine < 10 μm thick, no small echinae on top of muri: 12

12 a   Colpi short and porus-like, apocolpium large. Columellae branching and anastomosing to form the reticulum. Reticulum is partially obscured by clavae-like projections on top of the muri (careful 'LO' analysis needed. See scanning electron micrographs in Echlin, 1968): *Geranium*

12 b   Colpi long, apocolpium medium. Muri supported by one row of unbranched columellae. No projections on top of the muri: *Armeria* type, B-line

     [Includes *Armeria* and *Limonium*. See above for explanation of pollen dimorphism in the *Plumbaginaceae*.]

13 a   Grain circular in polar and equatorial view. Colpi short and narrow, inrolled except at the equator: *Fraxinus*

13 b   Grain circular or elliptic in equatorial view. Colpi long, narrow and inrolled or wide and boat-shaped: 14

14 a   Each colpus with a margo, i.e. the widths of the lumina get progressively smaller towards the colpus edge. The very edge of the colpus is tectate but this area is often inrolled: *Salix*

     [*Saxifraga stellaris* may key out here if the porus is not apparent.]

14 b   No margo around each colpus, i.e. lumina width remains approximately the same right up to the colpus edge (which is not tectate): 15

15 a Muri narrower than lumina, columellae longest in the centre of the mesocolpium, becoming shorter towards the poles and the edges of the colpi: *Sinapis* type
       [This type includes over twenty genera—see Erdtman, Praglowski and Nilsson (1963) for the full list and attempts at further division.]
15 b Muri as wide or wider than the lumina, columellae the same length all over the grain surface, or slightly longer at poles: 16
16 a Muri mostly duplicolumellate, colpus without granular membrane, grain < 25 μm in size: *Marrubium*
16 b Muri all simplicolumellate, columellae not arranged in a reticulate pattern, but evenly distributed. Colpus often boat-shaped with a granular membrane. Grain > 25 μm in size: 17
17 a Columellae coarse, often non-cylindrical. Muri with minute echinae on the surface. Colpus membrane with bacula as long as columellae on rest of exine: *Glaucium*
17 b Columellae finer, cylindrical; no echinae on muri. Colpus membrane with small granules. Grains circular in polar and equatorial view: *Helleborus, Chelidonium*
       [Comparison with type slides necessary for separation of these two types.]

Trizonocolpate: Echinate (Plates 24 and 25)

1 a Grain triangular-convex in polar view. Colpi short and apocolpium large in relation to grain size, grain thus broader than long. Two nexinous ridges (costae) cross each colpus, one on either side of the equator: 2
1 b Grains circular or lobed in polar view. Colpi long or short, but grain never broader than long. No nexinous thickenings crossing the colpi: 3
2 a Grain > 55 μm. Echinae > 1 μm long, rather sparsely spaced on the exine: *Lonicera*
2 b Grain < 55 μm. Echinae $\leqslant$ 1 μm long and densely clothing the surface of the grain: *Linnaea*
3 a Grains > 55 μm long. Exine > 5 μm thick (not including processes): 4
3 b Grains < 55 μm long. Exine < 5 μm thick (not including processes): 5
4 a Grain circular in both polar and equatorial view. Sexine of similar thickness all over the grain. Columellae thick (1–1·5 μm): *Succisa*
4 b Grain circular in polar view, elliptic in equatorial view. Sexine thickest and columellae longest in the apocolpium. Columellae slightly thinner ($\sim$ 1 μm): *Scabiosa*

5 a Echinae > 1 μm long, rather sparsely spaced: 6
5 b Echinae $\leqslant$ 1 μm long, sparsely or densely spaced: 7
6 a Echinae constricted at their bases and cylindrical for most of their length: *Rubus chamaemorus*
6 b Echinae not constricted at their bases, each one being conical and each set upon a low verruca: *Valeriana*
7 a Exine thickest, and columellae longest in the apocolpium: *Valerianella*
7 b Exine of the same thickness and columellae of the same length all over the grain surface: 8
8 a Faint suggestion of lines joining the bases of the echinae (use phase contrast if possible), i.e. infrareticulate: *Papaver*
       [e.g. *P. rhoeas*, *P. somniferum*, but see also page 60.]
8 b No suggestion of lines joining the bases of the echinae (use phase contrast if possible). Columellae very thin and densely packed: *Caltha* type
       [Includes *Caltha, Adonis, Aquilegia* and *Delphinium*.]

Trizonocolpate: Scabrate-Verrucate (Plates 25 to 27)

1 a With regularly spaced verrucae or microverrucae. On focusing down, small columellae are visible in the areas between the verrucae (phase contrast is useful here): 2
1 b Never with regularly spaced processes between which columellae can be seen in phase contrast (if processes regularly spaced then columellae not visible between them, if processes irregularly spaced then columellae visible between them): 5
2 a Columellae thin, of uniform thickness and length, visible as a dense carpet underneath the tectum which bears the verrucae: 3
2 b Columellae may vary in thickness and length, never forming a dense carpet but arranged in various patterns: 4
3 a Grains elliptic in equatorial view, tectum conspicuously thickened in the area just outside the apocolpium: *Teucrium*
3 b Grains circular in equatorial view, tectum not thickened just outside the apocolpium (i.e. same thickness all over grain): *Caltha type*
       [Includes *Caltha, Adonis, Aquilegia* and *Delphinium*.]
4 a Columellae nearly uniform in thickness and arranged in a reticulum underneath the tectum and verrucae (i.e. the grain is infrareticulate). The verrucae occur over the 'connections' of the reticulum: *Papaver*

       [See also above.]

4 b Columellae varying in thickness, sparse thick columellae surrounded by a light ring and then a carpet of fine, dense columellae (the light ring is caused by the absence of fine columellae in the immediate vicinity of the thick columellae): *Ranunculus* type

[Includes *Ranunculus, Clematis, Pulsatilla* and *Actaea*.]

5 a Processes are a mixture of scabrae, verrucae and sometimes rugulae irregularly scattered over the tectum. In phase contrast a fine, dense carpet of columellae is visible under the tectum: *Quercus*

[See scanning electron micrograph in Smit, 1973.]

5 b With verrucae and scabrae but no sideways-elongate elements (rugulae). Columellae either all coarse, or dimorphic (coarse and fine mixed): 6

6 a Columellae dimorphic, sparse large columellae surrounded by a light ring caused by the absence of the fine columellae in the immediate vicinity of the large columellae: *Ranunculus* type

[Includes *Ranunculus, Clematis, Pulsatilla* and *Actaea*.]

6 b Columellae monomorphic, all fairly coarse: 7

7 a Grains tectate perforate (examine carefully under × 1000. Phase contrast and an optical section are useful in identifying the tectum). Small verrucae or scabrae scattered over the tectum, their distribution bearing no relation to that of the columellae. Grains ± operculate. If operculate then operculum has same exine pattern as mesocolpia: *Saxifraga granulata* type

[Includes *S. granulata, S. hypnoides, S. cespitosa, S. rosacea*, and maybe *S. tridactylites* and *S. hirculus*, see the descriptions in Ferguson and Webb, 1970. For further descriptions of this type see Faegri and Iversen, 1964; Birks, 1973a]

7 b Grain verrucate, verrucae corresponding with columellae. Grain never operculate: 8

8 a Grain tectate perforate, with the perforations and small echinae in between the verrucae (use phase contrast if available). Columellae and verrucae rather densely packed. Exine adjacent to colpus the same thickness as exine in the middle of the mesocolpium. Consult a type slide: *Spergula* type

[Includes *Spergula* and *Spergularia*.]

8 b Grain tectate (phase contrast). Columellae and verrucae less densely packed. Exine adjacent to colpus thicker than exine in the middle of the mesocolpium. Consult a type slide: *Anemone*

Trizonocolpate: *Rugulate-Striate* (Plates 27 and 28)

1 a Striations very fine and faint, clearly visible only under × 1000 and phase contrast. Striations all running in an equatorial direction, i.e. at right angles to the colpus. Colpus granulate: n-*Trollius*

1 b Striations stronger, running either equatorially or meridionally (parallel to colpus) or in many directions: 2

2 a Muri each with a 'beaded' appearance at highest focus due to the columellae underlying it. No bacula visible between the muri: 3

2 b Muri very clear cut without any such beaded appearance at highest focus because each murus may be supported by more than one row of columellae. Bacula may or may not be visible between the muri: 4

3 a Muri long, of similar width, rarely branched, parallel to one another and running predominantly in a meridional direction. Grain elliptic in equatorial view: *Acer*

3 a Muri long, of similar width, rarely branched, par-branched and worm-like (i.e. grain tending towards rugulate). Grain rhombic-obtuse in equatorial view: *Dryas*

[See also Rosaceae, page 71.]

4 a Columellae minute and difficult to discern (view at × 1000). Grain tectate. Up to three rows of columellae support one murus. Muri straight, running meridionally in the mesocolpium and forming intercrossing groups in the apocolpium: *Menyanthes*

4 b Columellae thicker (discernible easily at × 1000). Grain semitectate. One or two rows of columellae per murus. Muri always curved, anastomosing or running in wide swirls which resemble the pattern of a fingerprint. Pattern in the mesocolpium not markedly different from that at the apocolpium: *Saxifraga oppositifolia* type

[Includes *S. oppositifolia, S. aizoides*. See Ferguson and Webb, 1970; Ferguson, 1972, for detailed descriptions and criteria for separation of these and other species.]

Trizonocolpate: *Baculate* (Plate 29)

1 a Bacula all of uniform height and very densely packed so that their heads appear polyhedral in surface view. Bacula may vary in thickness: 2

1 b Bacula may or may not be uniform in height, but always sparsely distributed over the surface of the grain (i.e. not so densely packed as to appear polyhedral): 4

2 a Bacula differing in width, some being two or

three times as wide as adjacent ones: *Linum bienne* type

> [Includes *L. bienne* and *L. usitatissimum*. For a discussion of the pollen morphology in *Linum* see Saad, 1961.]

2 b Bacula all the same width: 3

3 a Number of bacula across a mesocolpium from one colpus edge to the other is ten or less. Grain small: *Radiola*

3 b Number of bacula across a mesocolpium is > 20. Grain larger: *Linum catharticum*

4 a Grain circular in equatorial view, layer of densely packed small bacula between the bases of the large ones: *Viscum*

4 b Grains elliptic in equatorial view, exine tectate between bases of bacula. Bacula constricted at their bases: *Rubus chamaemorus*

Trizonocolpate: Clavate (Plate 29)

1 a With fairly densely packed clavae, very variable in height and width of head. Clavae in centre of mesocolpium longest and with widest heads: *Ilex*

1 b With fairly sparse clavae, more uniform in height and width and with a layer of densely packed small bacula between the bases of the clavae. Colpi may have an equatorial constriction: *Viscum*

Tetrazonocolpate (Plate 30)

1 a Grains reticulate, colpi short (full length visible in equatorial view): 2

1 b Grains psilate or scabrate-verrucate. Colpi ± short: 3

2 a Grain circular in both polar and equatorial view, columellae coarse, only united to a reticulum at the very top: *Fraxinus*

2 b Grain rectangular obtuse to elliptic in both polar and equatorial views. Grain broader than long, columellae fine ± visible under the muri. Reticulum very thin-walled and delicate: *Impatiens*

3 a Colpi long, curving around grain, wide and boat-shaped. Grain ⩾ 30 μm in length: *Viola arvensis* type

> [Includes *V. arvensis*, *V. tricolor* and some grains of *V. palustris*, *V. riviniana* and others.]

3 b Colpi shorter (whole length visible in equatorial view), not curving around the grain. Colpi ± wide but always parallel-sided with rounded ends: 4

4 a Colpi wide, with edges very ill-defined and diffuse. Colpus membrane often present so that colpus is often difficult to differentiate from mesocolpium except by the difference in the thickness of the exine: *Hippuris*

4 b Colpi with well-defined ± thickened margins: 5

5 a Colpus thickenings protruding—about 3 μm thick: *Myriophyllum spicatum*

5 b Colpus with margins hardly thickened, and not protruding. Exine next to colpus about 2·5 μm thick: *Myriophyllum verticillatum*

Pentazonocolpate (Plate 30)

1 a Grains eureticulate, colpi long, narrow and cracklike, edges well-defined: *Primula vulgaris* type

> [Includes *P. vulgaris*, *P. veris* and *P. elatior*.]

1 b Grains scabrate, colpi short, ± narrow, edges ± well-defined: 2

2 a Colpi edges well-defined, colpi may be narrow but not cracklike: *Myriophyllum spicatum*

2 b Colpi edges not well-defined, being diffuse and ragged. Colpus membrane often in place so that it is difficult to distinguish colpus from mesocolpium. Colpi always wide: *Hippuris*

Hexazonocolpate (Plate 30)

1 a Grains psilate-scabrate to microreticulate: 2

1 b Grains reticulate (lumina > 1 μm in diameter): 3

2 a Colpi slightly sunken, mesocolpium convex, grains tectate, psilate-scabrate or perforate: *Galium* type

> [Includes *Galium*, *Asperula*, *Rubia* and *Sherardia*.]

2 b Colpi not sunken, but flat on the grain surface. Sculpture always eureticulate: *Primula vulgaris* type

> [Includes *P. vulgaris*, *P. veris* and *P. elatior*.]

3 a Grains suprareticulate, columellae in an even carpet under the reticulum. Reticulum best developed in the centre of the mesocolpia where the lumina are very wide and the muri narrow, ± winding: *Prunella* type

> [Includes *Prunella*, *Nepeta* and *Glechoma*.]

3 b Grains eureticulate to tectate-perforate, columellae sitting in a reticulate pattern under the muri: *Mentha* type

> [Includes *Mentha*, *Lycopus*, *Thymus*, *Origanum* and *Salvia*.]

Polyzonocolpate (Plate 30)

1 a   Grains eureticulate, colpi not sunken, but flush with the grain surface, mesocolpium hardly convex: *Primula vulgaris* type

[See page 62.]

1 b   Grains psilate or tectate-perforate to microreticulate, colpi sunken, mesocolpium more or less convex: 2

2 a   Grains psilate, columellae invisible, grains long and narrow, almost syncolpate at the poles: *Ephedra*

2 b   Grains tectate-perforate to microreticulate, grains only slightly longer than broad, not syncolpate at the poles, mesocolpium definitely convex: *Galium* type

[Includes *Galium*, *Asperula*, *Rubia* and *Sherardia*.]

Tetrapantocolpate, Pentapantocolpate, Hexapantocolpate

The only British genera which exhibit these aperture systems with any frequency are *Ranunculus* and *Spergula* (see page 61). These systems occur occasionally in grains which are normally trizonocolpate and can be identified by reference to that key.

Polypantocolpate (Plate 31)

1 a   Grain reticulate, with up to thirty colpi. Free bacula or clavae visible in the lumina: *Polygonum amphibium*

1 b   Grains psilate to scabrate-verrucate, with twelve or less colpi: 2

2 a   Columellae fine to invisible: *Corydalis*

2 b   At least some of columellae easily visible: 3

3 a   Columellae longest and coarsest in the middle of the mesocolpium, shortest and finest near the colpus edge—exine consequently varying in thickness: *Montia fontana* type

3 b   Columellae of similar thickness, i.e. no difference between the ones near the colpi and those in the middle of the mesocolpium: 4

4 a   Tectum often undulating in optical section, sculpturing irregularly scabrate-verrucate. Columellae of two types—sparse coarse ones surrounded by a dense carpet of fine ones: *Ranunculus* type

[Includes *Ranunculus*, *Clematis*, *Pulsatilla* and *Actaea*.]

4 b   Tectum psilate, perforate, not undulating in optical section of the wall. Columellae uniformly coarse: *Spergula* type

N.B. Irregular polypantocolpate grains are found in some species of genera which in Britain are otherwise trizonocolpate or trizonocolporate e.g. *Linum anglicum*, *Papaver argemone*, *Rorippa sylvestris*.

Heterocolpate (Plates 31 to 33)

1 a   With three porate colpi and six non-porate colpi. The non-porate colpi situated one on either side of a porate colpus. Here the pori are actually short, transverse endocolpi covered by a sexinous equatorial bridge: 2

1 b   With three porate colpi and three non-porate colpi. Non-porate colpi situated exactly between porate colpi. Pori well-defined, never represented by equatorial bridges: 3

2 a   Grain coarsely rugulate in the apocolpia, psilate-scabrate in the mesocolpia. Endocolpus meridionally constricted: *Anthyllis*

2 b   Grain scabrate-rugulate in the apocolpia and mesocolpia. Endocolpus not meridionally constricted: *Verbena*

3 a   Pori are nexine features only (endopori), each equatorially elongated into a neat ellipse or obtuse rectangle. Grains often equatorially constricted (dumb-bell shaped), size < 14 μm: *Myosotis* type

[Includes *Myosotis*, *Mertensia*, *Cynoglossum* and some species of *Lithospermum*.]

3 b   Pori circular, features of sexine and nexine, grains never equatorially constricted, size > 14 μm: 4

4 a   Non-porate colpi much broader than porate colpi: *Peplis*

4 b   Non-porate colpi approximately as broad as porate colpi: *Lythrum*

Trizonocolporate

Subdivision of this group is on the basis of sculpture (see page 41)

PSILATE or SCABRATE-VERRUCATE (page 63)

RETICULATE (page 66)

ECHINATE or ECHINATE-VERRUCATE (page 70)

STRIATE or RUGULATE-STRIATE (p. 71)

BACULATE or PILATE (page 73)

Trizonocolporate: Psilate or Scabrate-Verrucate (Plates 31 to 36)

1 a   Grain approximately as long as broad: 2

1 b   Grain distinctly longer than broad: 24

2 a   Mesocolpium flat or even concave: 3

2 b   Mesocolpium convex or bulging: 12
3 a   With distinct, circular pori, sexine psilate, columellae invisible: 4
3 b   With each porus represented by either a bulging equatorial bridge, constriction or rupture to the colpus or by a transverse endocolpus; sexine psilate or scabrate-verrucate, columellae ± visible: 5
4 a   Porus small in comparison to grain size. Colpi narrow, parallel-sided, much less than half the width of the mesocolpium at the equator, curving in a characteristic manner near the porus. Grain circular to elliptic in equatorial view: *Frangula*
4 b   Porus large in comparison to grain size. Colpi wide, up to half the width of the mesocolpium at the equator. Colpi travel through an angle at the equator thus grain circular to rhombic in equatorial view: *Umbilicus*
5 a   With each porus an endoporus equatorially elongated into a rectangle or represented by a narrow transverse endocolpus: 6
5 b   Porus not equatorially elongated, being represented by a bulging of the exine at the equator, forming a bridge or constriction to the colpus (± ruptured): 8
6 a   Each endoporus a neat rectangle; columellae coarsest, longest and most widely spaced at the poles of the grain: *Hydrocotyle*
6 b   With a thin transverse endocolpus; columellae all fine and evenly distributed all over the grain surface or invisible: 7
7 a   Grain < 17 μm, columellae invisible, grain psilate: *Solanum dulcamara*
7 b   Grain > 17 μm, columellae ± visible, grain scabrate: *Solanum nigrum*
8 a   Grain > 35 μm in size, porus represented by a ± ruptured equatorial bridge or constriction to the colpus: 9
8 b   Grain < 35 μm, porus represented by a bulging of the exine near the colpus at the equator, or by a rupture which may be meridionally elongate: 10
9 a   Grain microscabrate-microverrucate all over, up to 60 μm in size, no fissure-like furrows on either side of each colpus: *Thelycrania sanguinea*
9 b   Grain psilate in the middle of the mesocolpium, irregularly verrucate-rugulate at the poles. Grain < 40 μm, a fissure-like furrow on either side of each colpus and running parallel to it: *Anthyllis*
10 a Columellae invisible, grain microscabrate-microverrucate: *Chamaepericlymenum suecicum*
                    [Cf. *Hippophaë*, page 65.]

10 b Columellae visible (at least in optical section), tectum psilate: 11
11 a Columellae longest and coarsest (i.e. exine thickest) at the poles of the grain. In equatorial view the exine 'pouts' at the apertures making the grain neatly rhombic-obtuse in shape: *Pleurospermum austraicum*
11 b Columellae the same length (therefore exine of uniform thickness) all over the grain. Equatorial exine not pouting and grain not so neatly rhombic obtuse. If porus is ruptured then its shape is that of a meridionally elongated rectangle: *Bupleurum* type
            [See the rest of the Umbelliferae on page 65]
12 a With porus equatorially elongated into a transverse endocolpus. Grain with or without operculate colpus: 13
12 b With approximately circular porus, often nexinously thickened and with or without protruding lips: 18
13 a Grain circular in polar and equatorial view, colpus operculate, operculum and edge of colpus bulging at the equator: *Poterium sanguisorba*
13 b Grain circular to lobed in polar view, rectangular obtuse to rhombic obtuse in equatorial view. Colpus a narrow fissure, not operculate: 14
14 a Endocolpus narrowly rectangular-obtuse. Exine the same thickness all over the grain surface, colpi travelling through an angle at the equator. Lips of exine project over porus at equator: *Filipendula*
14 b Endocolpus wider and not rectangular-obtuse. Exine varying in thickness over grain surface, colpi straight or gently curving: 15
15 a Exine thickest, and columellae longest at the poles of the grain: 16
15 b Exine thinnest at the poles of the grain. Columellae ± visible. Prominent costae along the colpi: 17
16 a Grain rectangular-obtuse in equatorial view, endocolpus elliptic-acute in shape, adjacent endocolpi either not meeting to form a girdle, or if just meeting then girdle widest under the colpi: *Polygonum aviculare*

N.B. *P. bistorta* type may key out here if the porus is slightly elongate, see page 65.

16 b Grain rhombic-obtuse in equatorial view, adjacent endocolpi nearly always fused so that a complete equatorial girdle is formed: *Polygonum convolvulus* type

            [Includes *P. convolvulus* and *P. dumetorum*.]

17 a Grain large, > 30 μm, columellae visible: *Lycopsis*

17 b Grain small, < 12 μm, columellae invisible: *Cynoglossum*

18 a Exine varying in thickness over the surface of the grain: 19

18 b Exine of uniform thickness all over the grain: 20

19 a Exine thickest and columellae longest in the middle of the mesocolpium. Grain lobed in polar view: *Artemisia*

[Includes *A. vulgaris, A. norvegica, A. maritima, A. absinthium* and *A. campestris.* See also page 70.]

19 b Exine thickest at the poles of the grain. There can be two types of exine structure: one where the columellae are distinctly visible and the other where the exine appears solid and without structure. If columellae are visible then they are longest and coarsest at the poles. Grain circular in polar view: *Polygonum bistorta* type

[Includes *P. bistorta* and *P. viviparum.*]

20 a Colpi wide and boat-shaped (up to half the width of the mesocolpium), grain completely psilate, columellae invisible. Grain < 20 μm in size: Crassulaceae, *Umbilicus* type

(Includes *Tillaea* and *Umbilicus.*]

20 b Colpi narrow, fissure-like, grain granulate to scabrate verrucate, columellae ± visible. Grain > 20 μm: 21

21 a With portions of exine projecting like lips or a beak at the equator: 22

21 b No such projecting portions of exine, porus all endoporus covered, by a flat layer of sexine: 23

22 a Colpi long, therefore apocolpium small: *Hippophäe*

22 b Colpi short, therefore apocolpium large: *Ludwigia*

[See also *Medicago*, page 66.)

23 a Colpi short (whole length visible in equatorial view) apocolpium large, endoporus large—up to 6 μm in diameter: *Fagus*

23 b Colpi long (whole length not visible in equatorial view), apocolpium consequently smaller. Endoporus smaller, < 5 μm in diameter: *Oxyria* type

[Includes all spp. mentioned under this type on page 68, plus some grains of *Rumex acetosella*.]

24 a Porus an endoporus, equatorially elongated into a neat rectangle, or parallel-sided transverse endocolpus or equatorial girdle: 25

24 b Porus a circular to elliptic endoporus or simply a bridge or constriction to the colpus, but not appreciably equatorially elongated (if slightly elongate then never neatly rectangular): 35

25 a Grains elliptic to rhombic-obtuse in equatorial view: 26

25 b Grains rectangular obtuse, ± concave in the equatorial region (dumb-bell shaped): 29

26 a Nearly always with a complete nexinous equatorial girdle, exine thickest and columellae longest at the poles: *Polygonum convolvulus* type

[Includes *P. convolvulus* and *P. dumetorum*.]

26 b With transverse endocolpi never quite anastomosing to form a girdle, exine the same thickness all over the grain, columellae invisible: 27

27 a Nexine twice as thick as the sexine, grains > 20 μm in size, transverse endocolpi ragged-ended, coming near to forming a girdle. Thick costae to the colpi: *Glaux*

27 b Nexine thin or stratification of exine indistinct, grains < 20 μm. Endocolpus extending into the mesocolpium for a short distance only: 28

28 a Exine approximately 1 μm thick, grain > 12 μm: *Castanea*

28 b Exine < 1 μm thick, grain < 12 μm: *Cynoglossum*

29 a Exine thickest, columellae longest and most widely spaced at the poles of the grain: 30

29 b Exine the same thickness all over, or thickest down the centre of the mesocolpium and thinnest at the poles. Columellae ± visible: 31

30 a With elliptic-acute transverse endocolpus, grain never concave or constricted in the equatorial region: *Polygonum aviculare*

30 b With endoporus rectangular, rectangular-obtuse, or transverse endocolpus present (rarely elliptic-acute). Grains may be constricted or concave in the equatorial region: Umbelliferae 1

[e.g. *Heracleum, Sanicula, Smyrnium* and others. See Cerceau, 1959.]

31 a Exine thickest, and columellae longest at various regions in the mesocolpium, e.g. down the centre, near the colpi or towards the apocolpium. Grain consequently triangular-obtuse in polar view: 32

31 b Exine of the same thickness all over the grain, columellae fine to invisible. Grain circular in polar view: 33

32 a Grain with endocolpi fused into a complete nexinous equatorial girdle, costae bordering girdle projecting inwards. Columellae branched, tectum thick. Grain never concave at the equator: *Centaurea cyanus*

32 b Without a complete equatorial girdle, if transverse endocolpus has costae then they do not project inwards to the same extent. Columellae unbranched, tectum thin—may disappear

completely at the poles so that the apocolpium is pilate. Grain ± concave at the equator: Umbelliferae 2

(e.g. *Daucus, Myrrhis, Conopodium, Aethusa, Pastinaca, Scandix, Oenanthe, Angelica* and many others. See Cerceau (1959) for discussion on pollen types in the Umbelliferae.)

33 a Grains > 25 μm, with huge costae to the colpi (exine in costae area three or more times as thick as elsewhere on the grain): *Vicia cracca* type

[Includes *V. cracca, V. sepium, V. orobus* and *Lathyrus montanus*. See below.]

33 b Grains < 25 μm, with small costae to the colpi: 34

34 a Columellae fine to invisible (at × 1000 and phase contrast): *Lotus* type

[See also below.]

34 b Columellae visible (at × 1000 and phase contrast): Umbelliferae 3

(e.g. *Crithmum, Apium, Conium,* and perhaps *Berula* and *Cicuta*).

35 a Porus neatly circular or elliptic grains never rhombic obtuse: 36

35 b Porus represented by an equatorial bridge, constriction or irregular rupture to the colpus. Grains ± rhombic obtuse: 39

36 a Exine thickest and columellae (if visible) longest at the poles: *Polygonum bistorta* type

[See page 65.]

36 b Exine of the same thickness all over the grain: 37

37 a Colpus wide and boat-shaped, colpus membrane with coarse spines or verrucae. Porus not covered by a layer of sexine. Grain may be finely rugulate-striate: *Aesculus*

37 b Colpus narrow and fissure-like, no spines on the membrane. Porus a nexine feature only (i.e. endoporus), being covered by a thin layer of sexine: 38

38 a Endoporus circular or meridionally elongate, grain completely psilate (phase contrast): *Lathyrus palustris*

38 b Endoporus equatorially elongate, grain often with large undulating rugulae in the mesocolpium (phase contrast): *Vicia cracca* type

[Includes *V. cracca, V. sepium* and *Lathyrus montanus*.]

39 a Mesocolpium flat, grain finely scabrate-verrucate all over. Grain size > 30 μm: *Thelycrania sanguinea*

39 b Mesocolpium flat or convex, grain ± scabrate-verrucate ± 30 μm: 40

40 a Colpi narrow and crack-like, grain elliptic to rectangular-obtuse in equatorial view: 41

40 b Colpi wide, boat-shaped, grain elliptic in equatorial view: 42

41 a Grains < 25 μm, columellae fine to invisible, grain rectangular obtuse in equatorial view: *Lotus* type

[Includes *Lotus, Ornithopus* and *Tetragonolobus*. These can be separated by reference to type slides.]

41 b Grains > 25 μm, columellae fine: *Medicago sativa* type

[Includes *M. sativa* and *M. falcata*. See also page 67.]

42 a Columellae visible as a dense and even carpet. Grain tending towards tectate-perforate or microreticulate (× 1000). Grain < 30 μm: *Lobelia* type

[Includes *Lobelia dortmanna* and *Digitalis*. See also page 68.]

42 b Columellae fine, ± visible as a dense and even carpet. Sculpture tending towards minutely rugulate-striate (× 1000, phase contrast helps). Costae to the colpi sometimes visible: *Crataegus* type

[Includes *Crataegus, Sorbus aria, S. torminalis* and *Rubus saxatilis*. The rugulate striate pattern can be much coarser—see page 41.]

Viola palustris type

[Includes *V. palustris* and *V. riviniana*. See also page 62.]

[N.B. Refer to type slides for separation of *Crataegus* type and *Viola palustris* type.]

Trizonocolporate: Reticulate (Plates 37 to 43)

1 a Grains suprareticulate (page 43) or foveolate (with perforations in the tectum, width of perforations being ≤ width of intervening walls): 2

1 b Grains eureticulate, widths of lumina may or may not be > width of muri: 16

2 a Grain longer than broad, shape in equatorial view rectangular-obtuse to rhombic-obtuse, in polar view always circular: 3

2 b Grain as long as broad, shape in equatorial view circular to rhombic-obtuse, in polar view circular to triangular-obtuse: 11

3 a Exine > 3 μm thick, foveolate, columellae coarse, branching near the tectum (see optical section): 4

3 b Exine ≤ 3 μm thick, suprareticulate, columellae fine to invisible, never branched: 5

4 a Exine thickest and columellae longest at the poles of the grain. Columellae may also appear

coarser and more widely spaced at the poles: *Polygonum bistorta* type

> [Includes *P. bistorta* and *P. viviparum*.]

4 b Exine of the same thickness and columellae of the same length all over the grain, columellae coarse at bases but branching into many fine rods before meeting the tectum: *Fagopyrum* type

5 a Grain tending to be pear-shaped in equatorial view, i.e. broadest at one pole and narrowest at the other pole. Pori situated nearer to the broader pole than to the narrower pole, i.e. not in the equatorial region of the grain: *Echium*

5 b Grain rectangular-obtuse or rhombic-obtuse but never pear-shaped. Pori always situated in the equatorial region: 6

6 a More than fifteen lumina across a mesocolpium at the equator of the grain: 7

6 b Less than fifteen lumina across a mesocolpium at the equator of the grain: 10

7 a Grains with obvious reticulum in mesocolpium, but either no reticulum, or minute reticulum, or only a few pits, in the apocolpium. Columellae ± visible under × 1000 and phase contrast: 8

7 b Grains with reticulum more or less similar in mesocolpium and apocolpium; columellae fine to indistinct: 9

8 a Endoporus well-defined, neatly elliptic. Colpi with huge nexinous thickenings (costae). Exine in area of costae approximately three times as thick as the exine elsewhere on the grain: *Lathyrus*

> [Some species only, e.g. *L. pratensis*, *L. maritimus*, *L. nissiola*. See pages 66 and 67.]

8 b Without such a well-defined elliptic endoporus, often having only a slight constriction of the colpus in the equatorial region. No such huge costae to the colpi (exine in the area of the costae less than three times as thick as exine elsewhere on the grain: *Astragalus* type

> [Includes *Astragalus*, *Oxytropis* and a few *Trifolium* species.]

9 a Lumina in the centre of a mesocolpium slightly smaller in diameter than those towards the apolcolpia or the colpi. Grains always rectangular-obtuse in equatorial view, size > 20 μm: *Ononis* type

> [Includes *Ononis* and *Melilotus*. See Faegri and Iversen, 1974.]

9 b Lumina in the centre of a mesocolpium no smaller than those towards the apocolpia or the colpi. Grains ± rectangular obtuse, sometimes elliptic, size < 20 μm: *Hypericum perforatum* type

> [Includes *H. perforatum*, *H. tetrapterum*, *H. pulchrum*. See also *Glaux*, page 67.]

10 a Endoporus usually ill-defined, not neatly elliptic, but more a constriction or rupture of the colpus. Grain shape varies between rhombic-obtuse and rectangular-obtuse. Colpi may or may not have costae: *Trifolium* type

> [Includes most species of *Trifolium* and some of *Medicago*, e.g. *M. lupulina*.]

10 b Endoporus well-defined, neatly elliptic, surrounded by a ring-like costa. Very thick costae present along the margins of the colpi (exine in costa area two to three times as thick as the exine elsewhere on the grain). In equatorial view grain shape always rectangular-obtuse: *Vicia sylvatica* type

> [Includes *Vicia* and *Lathyrus*, e.g. *V. sylvatica*, *V. sativa*, *V. Sepium*, *L. sylvestris* and *L. montanus*.]

11 a Exine < 2 μm thick, columellae maybe indistinct at × 1000 (even in phase contrast): 12

11 b Exine > 2 μm, columellae visible at × 1000, especially so in phase contrast: 13

12 a Porus distinct, small and markedly equatorially elongated. Grain triangular-obtuse in polar view: *Rhamnus catharticus*

12 b Porus indistinct, consisting of a large irregularly ruptured area at the equator. Porus often elongated meridionally. Grains triangular-obtuse to circular in equatorial view: *Genista* type

> [Includes *Genista*, *Ulex*, *Cytisus* and *Sarothamnus*.]

13 a Colpi very short, apocolpium correspondingly large. Huge costae to the pori (three times as thick as the exine elsewhere). One columella directly under each funnel-shaped lumen (i.e. columella forms the 'stalk' of the funnel): *Tilia*

> [See Andrew, 1971.]

13 b Colpi long, no such huge costae to the pori: 14

14 a Exine thickest and columellae longest at the poles. Columellae appear branched underneath the tectum (a difficult feature, even in optical section). Porus well-defined, neatly elliptic: *Polygonum bistorta* type

> [Includes *P. bistorta* and *P. viviparum*. See also page 65.]

14 b Exine either the same thickness all over, or thickest in the middle of the mesocolpium: 15

15 a Columellae fine—muri being duplicolumellate and short columellae often present in the lumina. Colpi widening towards the equator: *Hypericum elodes*

15 b Columellae coarse—lumina only slightly wider than columellae. Each colpus narrow with a wide, granulate margo: *Euphorbia*

16 a Lumina of reticulum < 1 μm in width (microreticulate): 17

16 b   At least some lumina (e.g. those in the centre of the mesocolpium) > 1 μm in width: 30

17 a   Columellae indistinct to invisible in an optical section of the grain wall ($\times$ 1000). May or may not be visible in 'LO' analysis or in phase contrast: 18

17 b   Columellae visible in optical section of the grain wall ($\times$ 1000), visible in 'LO' analysis and in phase contrast: 22

18 a   With a thin, ragged-ended transverse endocolpus. Nexine much thicker than sexine. Muri narrower than lumina. Large costae to the colpi: *Glaux*

18 b   Porus represented by either a circular endoporus, a rupture, a constriction or bridge to the colpus, but no transverse endocolpus present. Nexine usually thinner than sexine. Muri $\pm$ as wide as the lumina: 19

19 a   Grains often almost syncolpate at the poles (colpi very long, therefore apocolpium very small). In equatorial view grain shape varies from circular to elliptic but never rectangular-obtuse. Colpi only faintly constricted at the equator: 20

19 b   Grains with apocolpium slightly larger and no tendency towards the syncolpate condition. Shape in equatorial view varies from circular to rectangular-obtuse. Colpi constricted or ruptured at equator and bordered by costae for the rest of their length: 21

20 a   Grain < 18 μm in length. Apocolpium with lumina so small as to appear tectate-perforate: *Chrysosplenium*

20 b   Grain > 18 μm in length. Apocolpium reticulate: *Saxifraga nivalis*

21 a   Grain as long as broad: *Elatine*

21 b   Grains always longer than broad: *Hypericum perforatum* type
[Includes *H. perforatum, H. pulchrum, H. tetrapterum* and others. See also *Echium*, page 67.]

22 a   Colpi narrow, parallel-sided $\pm$ fissure-like: 23

22 b   Colpi wide, boat-shaped, i.e. widest at equator and narrowing to pointed ends at the poles: 28

23 a   Transverse endocolpus present (careful 'LO' analysis is needed to see through the covering sexine). Sometimes adjacent endocolpi fuse so that a complete equatorial girdle is present: *Anagallis tenella* type
[Includes *A. tenella* and *Lysimachia nemorum*.]

23 b   Endoporus circular or elongated into a rectangle, but no fused transverse endocolpi present: 24

24 a   Porus represented by a rectangular endoporus or diffuse, ruptured area, Grain circular to triangular-obtuse in polar view. Colpi travelling

through an angle at equator, i.e. not curving gently along their length: 25

24 b   Endoporus circular, grain always circular in polar view. Colpi curving gently along their length: 26

25 a   Grain > 18 μm, endoporus elongated into a rectangle: *Trientalis*

25 b   Grain < 18 μm, porus a small, diffuse rupture: *Samolus*

26 a   Colpi short (full length usually visible in equatorial view) columellae sitting in a reticulate pattern, i.e. lying under the muri only, no columellae visible in the lumina. Muri nearly always simplicolumellate (phase contrast and $\times$ 1000 essential): *Rumex acetosa*

26 b   Colpi long (full length not always visible in equatorial view) Columellae either in a reticulate pattern, i.e. localized under the muri, or in a dense, even carpet under both muri and lumina: 27

27 a   Muri appear slightly winding and dupli- or pluricolumellate, columellae visible in the lumina (phase contrast and $\times$ 1000 essential): *Rumex acetosella* (*sensu lato*)
[But see also pages 69 and 73 as the pollen types within this aggregate are variable.]

27 b   Muri not winding and usually simplicolumellate. Rarely with columellae in the lumina (phase contrast and $\times$ 1000 essential): *Oxyria* type
[Includes *Oxyria, Rumex crispus, R. conglomeratus, R. maritimus, R. sanguineus, R. pulcher* and some grains of *R. obtusifolius* and *R. aquaticus*.]

28 a   Grain > 20 μm in length. Colpi up to 7 μm wide at equator: *Digitalis*

28 b   Grains < 20 μm in length: 29

29 a   Margins of the colpi tectate: *Linaria*

29 b   Grain reticulate right up to the edges of the colpi, although the lumina become smaller towards the colpi: *Saxifraga stellaris*

30 a   Colpi narrow, parallel-sided, often crack- or fissure-like: 31

30 b   Colpi wide, boat-shaped (i.e. widest at equator and narrowing to pointed ends towards the poles): 41

31 a   Porus represented by a slight equatorial bridge, rupture or constriction to the colpus: 32

31 b   Porus well-defined (circular, elliptic or equatorially elongated to a rectangle) or fused transverse endocolpi present: 35

32 a   Colpi short (whole length visible in equatorial view), apocolpium large. No margo present, i.e. lumina remain approximately the same size right up to the edges of the colpus. The colpi may be inrolled except at the equator: *Fraxinus*

32 b Colpi long (whole length no visible in equatorial view) apocolpium smaller. Each colpus with a margo, i.e. the lumina become smaller in size near the colpus edge: 33

33 a Lumina becoming so small near the colpus edge as to give a tectate margo (tectate region may be difficult to detect if colpus is inrolled): *Salix*

33 b Margo not tectate, reticulum continues right up to the colpus edge: 34

34 a Grain < 20 μm, circular in equatorial view, colpi travelling through an angle at the equator: *Hottonia*

34 b Grain > 20 μm, elliptic in equatorial view, colpi curving gently from pole to pole: *Sambucus*

35 a Pori elongated equatorially, or transverse endocolpi present (adjacent ones may fuse to form an equatorial girdle): 36

35 b Pori circular to slightly elliptic: 38

36 a Each porus elongated into an obtuse rectangle: *Hedera helix*

36 b Transverse endocolpi present, being either boat-shaped (i.e. with acute ends) or fused with the adjacent endocolpi to form an equatorial girdle: 37

37 a Lumina in centre of mesocolpium two to three times larger than those at the margins of the colpi: *Lysimachia vulgaris*

[Includes *L. vulgaris*, *L. nummularia* and *L. thyrsiflora*.]

37 b Lumina in the centre of the mesocolpium only just larger than those at the margins of the colpi: *Anagallis arvensis* type

38 a Each colpus with a margo—lumina in centre of mesocolpium three to four times as wide as those adjacent to the colpus. Colpi long, narrow and parallel-sided with a small strip of colpus membrane visible. Porus very well-defined, circular, not covered by sexine: *Parnassia palustris*

[Cf. *Gentiana verna*, below.]

38 b No margo to each colpus—lumina of approximately the same size all over the grain. Colpi very narrow and fissure-like—no colpus membrane visible. Porus a circular ± well-defined endoporus: 39

39 a Colpi short (full length usually visible in equatorial view). Columellae sitting in a reticulate pattern, i.e. lying only under the muri and no columellae visible in the lumina. Muri nearly always simplicolumellate (phase contrast and × 1000 essential): *Rumex acetosa*

39 b Colpi long (full length not always visible in equatorial view). Columellae either in a

reticulate pattern, i.e. localized under the muri, or in a dense, even carpet under both muri and lumina: 40

40 a Muri appear slightly winding and dupli- or pluricolumellate, columellae visible in the lumina (phase contrast and × 1000 essential): *Rumex acetosella* s.l.

[See also p. 68 and trizonocolporate, psilate to scabrate-verrucate, as the grains in this aggregate are variable.]

40 b Muri not winding and usually simplicolumellate. Rarely with columellae in the lumina (phase contrast and × 1000 essential) *Oxyria* type

[Includes *Oxyria*, *Rumex crispus*, *R. conglomeratus*, *R. maritimus*, *R. sanguineus*, *R. pulcher* and some grains of *R. obtusifolius* and *R. aquaticus*. *Tuberaria guttata* may key out here if its tendency towards striation is not obvious, see page 71.]

41 a With pori circular or slightly elliptic, with or without a neat edge: 42

41 b Each colpus with an equatorial constriction, rupture or short transverse endocolpus. If pori are present they are never circular and neat-edged: 46

42 a Muri broad—often nearly as wide as the lumina, columellae coarse, only united to a reticulum at the very top. Grain large (> 30 μm), exine thick (> 2·5 μm): *Gentianella*

42 b Muri narrower than the lumina (at least in the centre of the mesocolpium). Reticulum more solid. Grains may or may not be > 30 μm: 43

43 a Porus large in comparison to grain·size (porus up to 5 or 6 μm in diameter). Exine 3 μm thick in the centre of the mesocolpium, thinning quickly to ~1 μm at colpi margins. Reticulate patterns continues right up to colpus edge: *Euonymus*

43 b Porus smaller in comparison to grain size, exine usually < 3 μm, margins of colpus tectate-perforate or striate: 44

44 a Porus small, neatly rounded and nexinously thickened. Grain shape in equatorial view elliptic (broader than long). Colpi at equator < 3 μm wide, apocolpium very small: *Parnassia*

44 b Porus larger, not nexinously thickened, grain in equatorial view circular or elliptic (longer than broad). Colpi at equator > 3 μm wide: 45

45 a Margin of colpus tectate-perforate: *Gentiana verna*

[*Gentiana nivalis* is similar to *G. verna* but much larger.]

45 b Margin of colpus striate: *Blackstonia*

46 a Grain without solid muri, i.e. 'reticulum' effect created by columellae sitting in a reticulate

pattern, small granules visible on the floor of each lumen. Nexine as thick as sexine in middle of the mesocolpium: *Viburnum*

46 b Grain with solid muri joining the tops of the columellae, ± granules in the lumina. Nexine thinner than sexine: 47

47 a Reticulum with lumina of similar sizes across the mesocolpium and right up to the edge of the colpus: 48

47 b Reticulum with lumina much smaller near the colpus than in the centre of the mesocolpium, or margin of colpus tectate: 49

48 a Columellae coarse (1 to 1·5 μm in width). Muri less than lumina in width. Exine 3 μm thick in centre of mesocolpium, apocolpium large: *Ligustrum*

48 b Columellae fine (0·5 μm or less). Muri equal to lumina in width. Exine < 3 μm, apocolpium very small (about 3 μm in diameter). Grain often rhombic-obtuse in equatorial view: *Diapensia*

49 a Lumina size decreasing slowly from centre of mesocolpium to edge of colpus. Columellae distinct, muri simplicolumellate: *Scrophularia* type

   [Includes *Scrophularia* and *Verbascum*. N.B. *Gentiana nivalis* may key out here—see *G. verna*, page 69.]

49 b Lumina size decreasing sharply to the tectate margo. Columellae ± distinct, muri ± simplicolumellate: 50

50 a Grain > 28 μm, circular to elliptic in equatorial view. Tectate margo not inrolled, muri duplicolumellate, costae to the colpi: *Hypericum elodes*

50 b Grain > 28 μm, elliptic in equatorial view. Tectate margo inrolled, muri simplicolumellate, no costae to the colpi: *Salix*

N.B. *Vitis* may key out at this point, but it was not included when the key was devised due to a lack of type material. Descriptions of the grain are to be found in Faegri and Iversen (1974) and in Erdtman, Berglund and Praglowski (1961).

Trizonocolporate: Echinate or Echinate-Verrucate
(Plates 44 and 45)

1 a Protruberances < 1·5 μm in height, all tiny echinae: 2

1 b Protruberances > 1·5 μm in height, echinae or verrucae: 5

2 a Exine < 2 μm in the centre of the mesocolpium. Exine of uniform thickness all over the surface of the grain. Porus equatorially elongated so that it is rectangular in shape. Edge of mesocolpium protrudes over the porus at the equator: *Filipendula*

2 b Exine > 2 μm in the centre of the mesocolpium. Exine either varies in thickness over the surface of the grain or tectum has undulating appearance. Porus not a rectangle, no projecting mesocolpium: 3

3 a Tectum solid, columellae distinct and longest in the mesocolpium. Pori (when visible) are small, not elongated equatorially: *Artemisia*

   [Includes *A. vulgaris*, *A. campestris*, *A. maritima*, *A. norvegica* and *A. absinthium*. See page 65.]

3 b No solid tectum apparent. The echinae and the layer underneath them appear traversed by very fine rods. Connections (i.e. columellae) between the solid nexine and this structured, echinate tectum either rudimentary or absent. Transverse equatorial endocolpus present: 4

4 a Columellae absent, endocolpi forming a complete equatorial girdle, grain elliptic in equatorial view. Outline of grain appears undulating because the echinae are broad-based: *Centaurea nigra* type

   [Includes *C. nigra* and *C. nemoralis*.]

4 b Columellae rudimentary, projecting down from the structured sexine but not making contact with the nexine. Nexine thus almost completely separated from sexine. Endocolpi do not join to form a complete girdle, outline of grain not undulating. Exine thickest at the poles of the grain: *Centaurea scabiosa*

5 a Echinae solid, cylindrical and constricted at the point of attachment to the tectum: *Rubus chamaemorus*

5 b Echinae solid or traversed by small rods. Echinae triangular in shape and never constricted at the point of attachment to the tectum: 6 (Compositae, Tubuliflorae)

6 a Grains > 35 μm in size: 7

6 b Grains < 35 μm in size: 11

7 a Columellae visible underneath the tectum (tectum and echinae are traversed by rods which are much finer and more densely packed than the columellae): 8

7 b No columellae visible underneath the tectum (only the fine, densely packed rods of the tectum are visible): 9

8 a Columellae slanting slightly underneath the tectum at the point where the echinae are situated, thus giving a 'star' pattern to each echina on 'LO' analysis. Colpi fairly short, grain circular in equatorial view and with a large apocolpium: *Cirsium*

8 b Columellae perpendicular under the tectum at the point where the echinae are situated (i.e. the

same as elsewhere on the grain). Colpi longer, apocolpium correspondingly smaller: *Serratula* type

[Includes *Serratula, Saussurea, Arctium* and *Carlina.*]

9 a Echinae > 3 μm long: *Aster* type

[Includes *Aster, Tussilago, Petasites, Filago* and others. N.B. this group overlaps in size with the *Bidens* type which has the same exine structure.]

9 b Echinae ⩽ 3 μm long: 10

10 a Grain elliptic in equatorial view, continuous equatorial endocolpus forming a girdle to the grain: *Centaurea nigra* type

[Includes *C. nigra* and *C. nemoralis.*]

10 b Grain circular in equatorial view, no endocolpus around equator of grain: *Carduus*

11 a Coarse columellae visible under the structured tectum and echinae. (Columellae can be detected by 'LO' analysis and by an optical section): *Anthemis* type

[Includes *Anthemis, Achillea, Chrysanthemum, Matricaria* and *Tripleurospermum.*]

11 b No columellae visible under structured tectum and echinae. 'LO' analysis and optical section reveal only the fine rods traversing the tectum: *Bidens* type

[Includes *Bidens, Inula, Pulicaria, Eupatorium, Erigeron, Bellis, Senecio, Gnaphalium, Solidago, Filago, Antennaria.* See also *Aster* type, the grains of which are slightly bigger.]

Trizonocolporate: Striate or Rugulate-Striate (Plates 45 to 47)

1 a Grain with pori circular and well-defined: 2

1 b Grain with each porus represented by an equatorial bridge, constriction or irregular rupture, or transverse endocolpus present: 6

2 a Colpi long, narrow and crack-like, apocolpium small and grain acuminate-obtuse in equatorial view. Striae strong, running parallel to the colpi, occasionally grading into a reticulum. Grain > 30 μm in length: *Helianthemum*

[The grains in *Tuberaria guttata* and *Gertiana pneumoananthe* may be similar. Compare with type slides.]

2 b Colpi wide and boat-shaped. In equatorial view grain circular to elliptic but not acuminate-obtuse. Rugulae or striae ± parallel to colpus, grain may or may not be < 30 μm: 3

3 a Colpus membrane with coarse echinae or verrucae, grain commonly rectangular-obtuse in equatorial view. The striations are fine and faint, travelling at 90 degrees to the colpus in the equatorial region of the mesocolpium: *Aesculus*

3 b Colpus membrane without echinae, grain circular to elliptic or rhombic obtuse in equatorial view. Striations at the equator either parallel to the colpi or running in other directions: 4

4 a Grains < 24 μm, columellae indistinct to invisible. Distinctive rugulate-striate pattern present, which is formed by intercrossing groups of very straight muri (× 1000 and may need phase contrast): *Sedum*

[Includes *S. acre, S. rosea, S. villosum* and *S. anglicum.*]

4 b Grains > 24 μm, columellae distinct. Striations ± distinct, never forming intercrossing groups: 5

5 a Striae in the mesocolpium all running parallel to the colpi, exine of similar thickness all over the grain: *Gentiana pneumonanthe*

[May include some grains of *Cicendia.*]

5 b Striae fine, in sets which run in different directions to one another. Junction between sets sharp. Some sets may run parallel to the colpus. Exine noticeably thickest in the centre of the mesocolpium: *Centaurium*

[May include some grains of *Cicendia.*]

6 a Grains semitectate, muri simplicolumellate. Sculpturing becoming reticulate-striate in the apocolpia: *Atropa*

6 b Grains tectate, muri simplicolumellate or duplicolumellate. Sculpturing always striate or striate-rugulate in the apocolpia: 7

7 a Grain > 35 μm, columellae very fine. Grain striate. Muri straight and coarse with edges very sharply cut and width up to 1·5 μm. Muri running predominantly meridionally in the mesocolpium, but extensively diving under one another in the apocolpium: *Menyanthes*

[Cf. *Prunus.*]

7 b Grain may or may not be < 35 μm, columellae varying from fine to coarse. Coarsely or finely striate or striate-rugulate. Muri rarely wide and sharply cut but often narrow and with indistinct edges: 8

[Mostly Rosaceae.]

8 a Grains with an operculum to each colpus. Operculum bears the same sculpturing pattern as the rest of the grain surface: 9

8 b Grains without opercula to the colpi. Each colpus narrow and fissure-like (however, grains may still appear lobed in polar view): 10

9 a Sculpturing indistinctly rugulate-striate, edge of the mesocolpium protruding like a collar at the equator of each colpus. Colpi short, grain circular (not lobed) in polar view: *Poterium sanguisorba*

9 b Sculpturing coarsely striate, muri straight and running meridionally from pole to pole. Edge

of mesocolpium not protruding at the equator of each colpus. Colpi long; grain circular to rounded triangular in polar view: *Potentilla* type

> [Includes *Potentilla* and *Fragaria*.]

10 a   With equatorial bridge constricted in the middle, endoporus rectangular-obtuse or dumb-bell shaped. Sculpturing finely rugulate-striate. Apocolpia flattened, grain < 22 μm: *Hippocrepis* type

> [Includes *Hippocrepis* and *Coronilla*.]

10 b   Equatorial bridge ± present, no rectangular-obtuse or dumb-bell shaped endoporus. Sculpturing finely or coarsely striate or striate-rugulate. Apocolpia not flattened, grains may or may not be > 22 μm: 11

11 a   Grain > 35 μm long, with a distinct transverse endocolpus crossing each of the colpi. Sculpture consists of fine, dense striations which run equatorially (at 90 degrees to colpi): *Agrimonia*

11 b   Grains may or may not be < 35 μm, but without such distinct transverse endocolpi crossing the colpi. Striations not running equatorially: 12

12 a   Striate, or rugulate-striate pattern distinctly visible under × 400, grain may or may not have an equatorial bridge to each colpus (two bulges of mesocolpium meet over the colpus at the equator): 13

12 b   Striate or rugulate-striate pattern only clearly visible under × 1000. Grain may or may not have an equatorial bridge to each colpus, often porus is represented by an irregular rupture: 15

13 a   Striae coarse and straight, not curved or branched but long and running almost from pole to pole. Bridge to colpus very prominent: *Geum*

13 b   Striae not always straight and running from pole to pole but curved and branched, or running in wide swirls: 14

14 a   With conspicuously curved and branched striae or worm-like rugulae, muri varying in width. Grains never with an equatorial bridge to the colpus, columellae fine: *Dryas*

> [See also page 61.]

14 b   Striations curved and branched, ± worm-like, equatorial bridge to each colpus present, columellae coarse: *Prunus* type

> [Includes *Prunus* and *Rosa*. Some forms of *Prunus padus* may approach the *Dryas* sculpturing type.]

15 a   Grains with equatorial bridge to each colpus: *Rubus*

15 b   Grains with no equatorial bridge to each colpus, porus commonly represented by a very irregular rupture: *Crataegus* type

[This type represents a number of Rosaceae genera which may be distinguished individually by reference to type slides. It includes *Crataegus*, *Sorbus*, *Pyrus*, *Malus*, some species of *Prunus* and *Rosa*. For useful descriptions of pollen morphology in the Rosaceae see Reitsma, 1966.]

Trizonocolporate: Baculate or Pilate (Plate 47)

Grains rhombic-obtuse to elliptic in equatorial view, with transverse endocolpi crossing the colpi. Bacula or pila with heads ± joined laterally to form irregular rugulae or a reticulum: *Mercurialis*

Tetrazonocolporate (Plate 48)

1 a   Grains rectangular-obtuse in equatorial view. Porus a nexine feature, equatorially elongated into a transverse endocolpus or complete equatorial girdle. Rather thick costae to colpi: *Pulmonaria* type

> [Includes *Pulmonaria*, *Lycopsis* and *Lithospermum*.]

1 b   Grains circular or elliptic in equatorial view, porus either circular and a nexine feature (i.e. endoporus covered by sexine) or represented by a constriction or rupture to the colpus. No costae to colpi: 2

2 a   Colpi wide, with constriction or rupture to each, sculpture psilate-scabrate, columellae fine to indistinct. Grains > 35 μm in length: *Viola arvensis* type

> [Includes *V. arvensis*, *V. tricolor* and some grains of *V. riviniana*, *V. palustris* and others. See page 62.]

2 b   Colpi narrow, crack-like, with covered circular endoporus. Sculpture eureticulate to psilate-scabrate, columellae distinct: 3

3 a   Grains with short colpi (full length usually visible in equatorial view). Sculpture eureticulate with columellae in reticulate pattern under muri. Muri almost as wide as lumina. Muri always simplicolumellate with no columellae in the lumina (use phase contrast): *Rumex acetosa*

3 b   Grains with longer colpi (full length not usually visible in equatorial view). Sculpture reticulate or microreticulate to psilate-scabrate, columellae sitting in a reticulate pattern or in a ± even carpet: 4

4 a   Sculpture microreticulate or tectate-perforate (phase contrast). Muri not winding and usually simplicolumellate. Rarely with columellae in the lumina: *Rumex obtusifolius* type

> [Includes *R. obtusifolius*, *R. aquaticus*, *R. hydrolapathum* and others, and some grains of *R. acetosella*.]

4 b Sculpture reticulate, muri appearing slightly winding and dupli- or pluricolumellate; columellae visible in the lumina (phase contrast is necessary): *Rumex acetosella* s.l.

N.B. Tetrazonocolporate grains often occur in types which are normally trizonocolporate, e.g. in the Rosaceae, especially *Rubus* and *Rosa*.

Pentazonocolporate

1 a Grains > 50 μm in length, porus represented by an irregular constriction or rupture to each colpus and never elongated to a transverse endocolpus: *Viola arvensis* type
   [Includes *V. arvensis*, *V. tricolor* and others.]
1 b Grains < 50 μm in length, each ectocolpus with a transverse endocolpus which fuses with neighbouring endocolpi to form an equatorial girdle. Large costae to the colpi: *Pulmonaria*

Hexazonocolporate (Plate 48)

1 a Grain with a thick non-perforate tectum (see optical section): *Sanguisorba officinalis*
1 b Grains eureticulate to tectate-perforate: *Pinguicula*

N.B. Reitsma (1966) regards *Sanguisorba officinalis* as trizonocolporate with very wide opercula to the colpi.

Polyzonocolporate (Plate 48)

1 a Grains reticulate or microreticulate. Colpi constricted at the equator: *Pinguicula*
1 b Grains psilate to scabrate with neat pori or equatorial girdle of fused endocolpi: 2

2 a Apocolpium with circular or irregular depressions: *Polygala*
2 b Apocolpium without such depressions: 3
3 a More than ten colpi, grain circular in polar view, rhombic obtuse in equatorial view: *Utricularia*
3 b With eight colpi and eight neat elliptic pori, grain rectangular-obtuse in equatorial view: *Symphytum*

Tetrapantocolporate, Pentapantocolporate, Hexapantocolporate

1 a Each colpus crossed by a transverse endocolpus, grain usually with six colpi: *Polygonum raii*
1 b Pori circular, covered, endopori; grains can have anything between four and six colpi: 2
2 a Grains eureticulate, columellae sitting in a reticulate pattern under the muri. Muri simplicolumellate, no columellae in the lumina (phase contrast). No more than four colpi: *Rumex acetosa*
2 b Grains tectate-perforate to eureticulate, columellae either in a dense carpet or in a reticulate pattern, localized under the muri. Four to six colpi and pori: 3
3 a Grains reticulate, muri slightly winding. Dupli- to pluricolumellate and usually with free columellae in the lumina (phase contrast): *Rumex acetosella* s.l.
3 b Grains reticulate to tectate-perforate, muri not winding. Usually simplicolumellate and without columellae in the lumina (phase contrast). No more than four colpi: *Rumex*
   [Includes *R. obtusifolius*, *R. hydrolapathum*, *R. aquaticus* and other *Rumex* species and also some grains of *R. acetosella*.]

N.B. Pantocolporate grains may occasionally appear in species which are normally tricolporate, e.g. Rosaceae, Compositae subfamily Tubuliflorae.

Glossary

Acuminate-obtuse: Shape of a grain in equatorial view where the poles are extended but obtuse. See Fig. 4.7.

Annulus: Border to a porus which is produced by either a thickening or thinning of the sexine. For an example of the latter case see Caryophyllaceae. See page 38 and Fig. 4.6.

Apocolpium: The area at a pole of a zonocolpate grain delimited by the ends of the colpi. See Fig. 4.5.

Baculate: With bacula.

Baculum (pl. Bacula): Pillar or rod-like element always longer than broad and higher than 1 μm. Also used as a synonym for columella by other authors, e.g. Erdtman, Berglund and Praglowski (1961). See Fig. 4.8.

Clava (pl. Clavae): A projecting element which is higher than broad and tapering towards the base, i.e. club-shaped. Height greater than 1 μm.

Clavate: With clavae.

Colpate: With one or more colpi.

Colporate: With a colpus and porus combined in the same aperture, e.g. *Fagus*. See page 35.

Colpus (pl. Colpi): An elliptic aperture resembling a groove or furrow and with a length/breadth ratio greater than 2, e.g. *Quercus*. See page 35.

Colpus membrane: Thin, usually structureless layer of exine which covers a colpus in the living pollen grain and is the area through which water is lost or absorbed. This membrane is often lost during fossilization. See also Operculum.

Columellae: General term for small, rod-like elements, radially directed and forming the inner layer of the sexine. They are attached at their bases to the nexine and at their heads to the tectum (when present). On focusing down through the exine they appear as small black dots. See page 41 and Fig. 4.9.

Columellate: With columellae.

Costa (pl. Costae): A thickening of the nexine near an aperture. See Fig. 4.6; see also Annulus and Margo.

Distal face: That part of a spore which faces away from the centre of the tetrad during meiosis. See Fig. 4.5 and Proximal face.

Distal pole: That pole of a zonoaperturate grain which faces away from the centre of the tetrad during meiosis.

Dizonocolpate: With two colpi arranged in an equatorial zone.

Dizonoporate: With two pori arranged in an equatorial zone.

Duplicolumellate: With columellae in two rows under each murus.

Dyad: Two grains united. See Fig. 4.4.

Echinae: Any sharply pointed sculpturing elements. They may vary from cylindrical to cone-shaped. See Fig. 4.8A.

Echinate: With echinae.

Ectoaperture: An aperture in the sexine.

Ectocolpus (pl. Ectocolpi): An elliptic ectoaperture with a length/breadth ratio higher than 2.

Ectoporus (pl. Ectopori): Circular or faintly elliptic ectoaperture with a length/breadth ratio smaller than 2.

Ektexine: Outer layer of exine which comprises tectum (if present), columellae and foot layer (nexine 1). It is equivalent to sexine plus nexine 1.

Endexine: Inner, sometimes laminated layer of the exine, synonymous with nexine 2 and underlying nexine 1 (foot layer).

Endoaperture: An aperture in the nexine.

Endocolpus (pl. Endocolpi): A colpus, i.e. an elliptic endoaperture with a length/breadth ratio greater than 2. The endocolpus may be covered by a layer of sexine. Endocolpi often cross ectocolpi at 90 degrees, and adjacent ones may fuse to form a girdle in zonoaperturate grains, e.g. *Centaurea cyanus*, Plate 31. See Fig. 4.6.

Endoporus (pl. Endopori): Circular or faintly elliptic endoaperture with a length/breadth ratio smaller than 2.

Equatorial bridge: Structure formed by two bulges of mesocolpium meeting over the colpus at the equator, e.g. *Geum*.

Equatorial view: View of a zonoaperturate grain where the equatorial plane is directed towards the observer. See Fig. 4.5.

Eureticulate: With partially dissolved tectum, i.e. heads of columellae joined in only one or two directions to form the muri of the reticulum. Distribution of columellae often corresponds to that of the muri although there may be free columellae in the lumina, e.g. *Salix, Ligustrum*. See Fig. 4.8A and Supra reticulate.

Fenestrate: Term given to grains of the Compositae subfamily Liguliflorae. These are either trizonocolporate or trizonoporate but these aperture systems are obscured by the very unusual sexine pattern of these grains. There are large 'gaps' (lacunae) in the sexine which are in a fixed geometrical pattern and are separated by high echinate ridges. See *Taraxacum*, Plate 2.

Foveolae: Large perforations in a tectum (i.e. > 1 μm in diameter). Perforations always less wide than the area between the perforations. Bacula may sometimes be seen in the foveolae. See Fig. 4.8A.

Foveolate: With foveolae.

Gemma (pl. **Gemmae**): Sculpturing element with width approximately the same as height and constricted at its base. Height always greater than 1 μm.

Gemmate: With gemmae.

Granules: General term for very small (< 1 μm high) structures on a grain surface that are not assignable to gemmae, bacula, echinae or clavae, e.g. *Populus*.

Heterocolpate: With some apertures colpi and some apertures colpi + pori.

Hexapantocolpate: With six colpi scattered all over the surface of the grain.

Hexapantocolporate: With six colpi scattered all over the surface of the grain, each colpus with a porus in its centre.

Hexapantoporate: With six pori scattered all over the surface of the grain.

Hexazonocolpate: With six colpi arranged in an equatorial zone.

Hexazonocolporate: With six colpi arranged in an equatorial zone, each colpus having a porus in its centre.

Hexazonoporate: With six pori arranged in an equatorial zone.

Inaperturate: Without apertures (pori or colpi).

Infrareticulate: With a reticulate pattern produced by the distribution of the columellae underneath a complete or partially dissolved tectum, e.g. *Papaver*.

Intectate: Sculpturing type in which the tectum is absent, i.e. the heads of the structural rods are free. Intectate grains can be baculate, echinate, clavate, gemmate, verrucate or granulate. See Fig. 4.8B and page 41.

Intine: The cellulosic innermost layer of the pollen grain wall which underlies the exine. It is the cell wall of the living pollen grain.

Lacuna (pl. **Lacunae**): Large gap in the sexine of fenestrate grains. The lacunae are separated by high sexinous ridges arranged in a fixed pattern. See page 43.

Lamellae: Tangential layers of exine material which are usually seen in the nexine.

'LO' pattern: Diffraction images produced during downward focusing through the exine. 'LO' is the term given to the pattern: bright islands and dark channels at high focus with dark islands and bright channels at low focus (from Latin *lux*, light and *obscuritas*, darkness). See Fig. 4.9.

Lumen (pl. **Lumina**): A gap or space between the walls of a reticulate, striate or rugulate sculpture.

Margo: Zone around a colpus formed by a sudden thinning or thickening of the sexine or by any other sexine structure different from the remaining sexine. See Fig. 4.6.

Meridional: Term applied to grain features (e.g. colpi) which run along lines joining the distal pole to the proximal pole over the grain surface.

Mesocolpium: The area of grain surface between two adjacent colpi. It is usually delimited by transverse lines drawn through the polar ends of the colpi. See Fig. 4.5.

Mesoporium: The area of grain surface between two adjacent pori. It is usually delimited by transverse lines drawn through the polar edges of the pori. See Fig. 4.5.

Microechinate: With minute echinae, less than 1 μm in height.

Microgemmate: With minute gemmae, less than 1 μm in height. See also Granules.

Microreticulate: With lumina of reticulum less than or equal to 1 μm in diameter.

Microrugulate: With minute rugulae, length less than or equal to 1 μm.

Microverrucate: With minute verrucae, less than 1 μm in height.

Monocolpate: With one colpus.

Monolete: Term given to pteridophyte spores with one aperture that resembles a groove or furrow.

Monoporate: With one porus only.

Murus (pl. **Muri**): A ridge or wall separating two lumina of reticulate, striate or rugulate sculpture.

Nexine: Inner unsculptured part of the exine which appears as a solid or occasionally laminated layer. See Fig. 4.2 and page 31.

'OL' Pattern: The reverse of the 'LO' pattern, i.e. dark islands and bright channels at high focus with bright islands and dark channels at low focus. See Fig. 4.9.

Operculate: With an operculum on some or all of the apertures.

Operculum: Thick membrane covering a porus or colpus. Generally there is a very thin area just inside the immediate margin of the aperture, so that the operculum is easily lost.

Optical section: View where focal plane is half-way through grain, i.e. thickness and structure of grain wall is visible.

Papillae: Hollow finger-like projections which are always longer than broad.

Pentapantocolpate: With five colpi scattered all over the surface of the grain.

Pentapantocolporate: With five colpi scattered all over the surface of the grain, each colpus having a porus in its centre.

Pentapantoporate: With five pori scattered all over the surface of the grain.

Pentazonocolpate: With five colpi arranged in an equatorial zone.

Pentazonocolporate: With five colpi arranged in an equatorial zone, each colpus having a porus in its centre.

Pentazonoporate: With five pori arranged in an equatorial zone.

Perforate: With tectum pierced by small holes. Diameter of holes less than 1 μm. See also Foveolate.

Perine: The outermost part of the exine in bryophytes and pteridophytes which may be more or less detached from the inner part.

Pilate: With pila.

Pilum (pl. Pila): Element consisting of a rod-like part (columella) and an apical swollen part (caput). See also Clava.

Pluricolumellate: With columellae in several rows beneath each murus.

Polar axis: A straight line connecting the distal and proximal pole of a pollen grain.

Polar view: View of a zonoaperturate grain where the polar axis is directed straight towards the observer. See Fig. 4.5.

Polyad: More than four grains united in a group. See Fig. 4.4.

Polypantocolpate: With more than six colpi scattered all over the surface of the grain.

Polypantoporate: With more than six pori scattered all over the surface of the grain.

Polyplicate: With more than six plicae (ridges) arranged in an equatorial zone. In practice this condition is almost indistinguishable from polyzonocolpate, and for this reason we have included *Ephedra* in the latter class.

Polyzonocolpate: With more than six colpi arranged in an equatorial zone.

Polyzonocolporate: With more than six colpi arranged in an equatorial zone, each colpus having a porus in its centre.

Porate: With one or more pori.

Porus (pl. Pori): A circular or slightly elliptic aperture with a length/breadth ratio smaller than 2.

Porus membrane: Thin, usually structureless layer of exine which covers a porus in the living pollen grain and is the area through which water is lost or absorbed. This membrane may be lost during fossilization. See also Operculum.

Proximal face: That part of a spore which faces towards the centre of the tetrad during meiosis. See Fig. 4.5. In trilete spores the three-slit aperture is on the proximal face.

Proximal pole: That pole of a zonoaperturate grain which faces towards the centre of the tetrad during meiosis.

Psilate: With completely smooth, sculptureless surface.

Rectangular-obtuse: Shape of a grain in equatorial view where the grain resembles a rectangle with the corners rounded. See Fig. 4.7.

Reticulate: With a reticulum.

Reticulum: A network or mesh-like pattern.

Rhombic-obtuse: Shape of a grain in equatorial view where the grain resembles a rhomboid with the corners rounded. See Fig. 4.7.

Rugulate: With sculpturing elements elongated sideways, length greater than twice breadth (i.e. muri). Elements very irregularly distributed. See Fig. 4.8A.

Saccate: With two sacci.

Sacci: Large hollow projections from the main body of the grain or spore. In some conifers the sacci are thought to be formed by 'sexine' and 'nexine' being pushed apart at the lower end of the columella-like structures, and the sexine then being very much expanded and stretched, e.g. *Pinus*.

Scabrae: Very small (< 1 μm high), isodiametric sculpturing elements, often resembling minute flakes or lumps, e.g. *Thelycrania sanguinea*. See Fig. 4.8A.

Scabrate: With scabrae.

Semitectate: Sculpturing type in which the tectum is partially absent, i.e. the perforations in the tectum are wider than or equal to the areas between the holes. Semitectate grains can be striate, reticulate or rugulate. See Fig. 4.8A and page 43.

Sexine: Outer sculptured part of the exine which in Angiosperms takes the form of a set of radially directed rods (columellae) supporting a roof (tectum). The tectum may be complete, partially dissolved or completely lacking. See Fig. 4.2 and page 31.

Simplicolumellate: With columellae in one row under each murus.

Striate: With a pattern consisting of approximately parallel muri and lumina. See Fig. 4.8A.

Suprareticulate: With a reticulum on top of the tectum. Pattern is thus independent of the distribution of the columellae, e.g. *Galeopsis*. Columellae are not always visible, e.g. in some members of the Papilionaceae. See Fig. 4.8A.

Syncolpate: With two or more colpi fused at their ends. See page 36.

Tectate: With a complete roof layer joining the heads of the columellae. See Fig. 4.8A.

Tectum: 'Roof' layer which joins the heads of the columellae and thus forms the outer layer of the sexine. It may be more or less dissolved. See Semitectate and Fig. 4.8A.

Tetrad: Four grains united in a group. See Fig. 4.4.

Tetragonal tetrad: Tetrad with the constituent grains arranged so that their centres are at the corners of a tetrahedron. See Fig. 4.5. ($\geqslant$ Tetrahedral tetrad).

Tetrapantocolpate: With four colpi scattered all over the surface of the grain.

Tetrazonocolpate: With four colpi arranged in an equatorial zone.

Tetrazonocolporate: With four colpi arranged in an equatorial zone, each colpus having a porus in its centre.

Tetrazonoporate: With four pori arranged in an equatorial zone.

Trilete: A three-slit aperture, appearing like a Y. This aperture is on the proximal face of a spore.

Trizonocolpate: With three colpi arranged in an equatorial zone.

Trizonocolporate: With three colpi arranged in an equatorial zone, each colpus having a porus in its centre (porus may be well defined and circular or represented merely by a bridge or constriction in the middle of the colpus).

Trizonoporate: With three pori arranged in an equatorial zone.

Verrucae: Wart-like processes, usually broader than high and never constricted at the base (i.e. nearly hemispherical), e.g. *Plantago major*. See Fig. 4.8A.

Verrucate: Possessing verrucae on the surface of the grain.

Vestibulate: With a vestibulum to each porus.

Vestibulum: Small chamber formed by the sexine and nexine splitting apart from one another in the vicinity of a porus. The chamber thus communicates with the inside and outside of the grain, e.g. *Betula, Alnus*. See page 38 and Fig. 4.6.

Chapter 6

Pollen Counting and Pollen Diagram Construction

Sampling and counting

The process leading up to the preparation of slides containing concentrations of pollen is essentially one of sequential sampling and subsampling. Each sampling stage involves a successive approximation and hence a decrease in the accuracy of the ultimate representation (see Woodhead and Hodgson, 1935).

Approximation takes place at the following stages:

1. A single borehole is taken to be representative of what is usually a considerable body of sediment. Even if several borings are subjected to pollen analysis, these samples represent a minute quantity of material when considered in relation to the total deposit.
2. In most analyses samples of sediment are taken from the core at selected intervals. In effect this is subsampling from the cross-sectional area of the core. In certain circumstances the entire cross-section may be used.
3. In most analyses this sample is treated chemically and a subsample is then mounted upon microscope slides. In certain absolute techniques the entire sample is mounted, but this is difficult, inefficient and time consuming.
4. Lastly, it is usual to count only a 'representative' sample of the grains upon a given slide. Again, certain absolute techniques require the counting of all grains upon a slide, but this is not normal practice.

In order that the final count of pollen grains should be as representative as possible of the true proportions of pollen types at that particular sediment depth, all efforts should be made to keep the sampling errors small. In practice this involves duplication at each sampling stage and counting a large number of pollen grains on the final slides.

Obviously, this process has feasibility limits in terms of the time taken in counting duplicate slides, and the duplication involved is subject to the law of diminishing returns, i.e. beyond a certain point the additional accuracy gained is not worth the effort expended. A final pollen count must therefore be determined which will most efficiently estimate the pollen proportions upon the sample slide.

The size of this count is difficult to determine because it will vary between different pollen types, depending on their frequency and the evenness of their dispersion on the slide. For example, Fig. 6.1 shows the relationship between the estimated proportions of two pollen types, Gramineae and Cyperaceae, with sample size on a slide in which the true proportions were of the order of 60% and 20% respectively. For the more abundant type, Gramineae, a count of only 50 total pollen grains gives a reasonable estimate of the true proportion of grass grains in the sample. However, in the case of Cyperaceae, which are three times less abundant, one needs a total pollen count of 150 in order to achieve a good estimate. Counts in excess of 150 total grains are wasteful of effort as far as these two pollen types are concerned, but may well be necessary if one wishes to have reliable estimates of the percentage frequency of scarcer grains.

In practice, the number of grains counted depends upon the diversity and especially the equitability (the way in which the grains present are distributed among the total pollen types present) of the pollen assemblage. In early work,

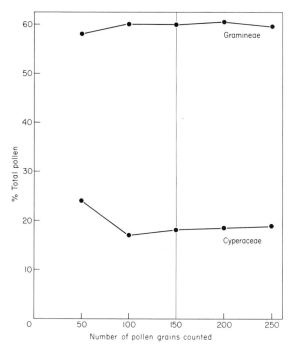

Fig. 6.1 Fluctuation of major pollen constituents with total pollen counted.

pollen analysis was used primarily as a tool for the investigation of forest history, hence the main concern was that major tree pollen constituents should be counted in sufficient number to estimate adequately their true proportions. It was found that 150 arboreal (tree) pollen grains provided an adequate sample and this sum became standard (Barkeley, 1934; Godwin, 1934). In more recent times the pollen analytical technique has been applied to wider problems of vegetational history and in these studies it has often been found valuable to use other methods for determining pollen sums (Wright and Patten, 1963) and proportional expressions.

The principles involved in determining the pollen count have already been outlined; the count should be sufficient to give a good approximation of the relative abundance of the pollen types which are of greatest interest to the study. It is this consideration which dictates the size of the pollen count. Where a study is aimed at the reconstruction of a regional environment which

was substantially wooded, yet where trees were not a major component of the local flora, the 150 arboreal pollen sum may be adequate. If certain minor components of the tree pollen sum, e.g. elm or lime, are of particular interest to the study, then it may be necessary to increase the total count to ensure adequate estimation of these types. Hence the total count must be determined by a consideration of the objects of the specific study and the adequate representation of those components which are of especial interest.

In some circumstances it may be possible to eliminate certain pollen types from the count after a representative sample of these has been counted. For example, in some ombrotrophic mire peats there may be a high proportion of Ericaceae and Cyperaceae pollen from species such as *Calluna*, *Erica* and *Eriophorum* which may have been growing locally. Often one is more interested in regional pollen which may be present in relatively small proportions. In such circumstances it is possible to assess the proportional relationship between the locally over-represented types and a non-local pollen type (or the arboreal pollen sum) by means of a fairly small sample count, e.g. less than 1000 total grains. When this has been established it is possible to continue the count ignoring these abundant but local pollen elements. In this way the total pollen count can be reduced without involving any significant loss of accuracy in the final proportions. Normally it is not necessary to apply such techniques in *Sphagnum* peats, but it is virtually essential in slower forming peats such as blanket peats (Moore and Chater, 1969a), which contain a larger proportion of angiosperm plants in their peat-forming vegetation.

Over-representation of a single pollen type in a percentage count is one of the reasons why certain absolute pollen counting methods are of interest. These involve calculation of the absolute density of pollen per unit volume or per unit weight of peat. A consideration of the nature of the sediment (i.e. its texture, specific composition, humification, compression, etc.) is essential for the interpretation of absolute pollen counts.

The factors influencing the rate of formation of a deposit were discussed in Chapter 2 where it became evident that although depth is a function of time, the relationship between the two may not be a simple linear one. Fig. 6.2 shows a hypothetical situation where three types of deposit have

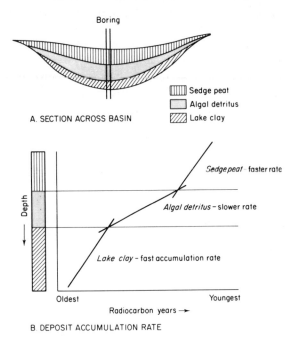

Fig. 6.2 Hypothetical relationship between depth, sediment type and age. In B the fluctuating rate is shown as a series of straight lines which are the best fit to the scatter of radiocarbon dates.

different accumulation rates (depth having been related to time by a series of radiocarbon dates). Suppose that algal detritus accumulates at a slower rate than sedge peat (the accumulation rate may vary even within one type of deposit). This means that it took more years to form 1 cm depth of the detritus than 1 cm depth of the sedge peat. If the rate of pollen input does not vary, 1 cm³ of algal detritus will contain more pollen than 1 cm³ of sedge peat. In fact 1 cm³ of deposit could contain from one to fifty or even more years of pollen rain per cm². This relationship of pollen/unit volume of deposit to the sediment type is shown in Fig. 6.3. Here the total

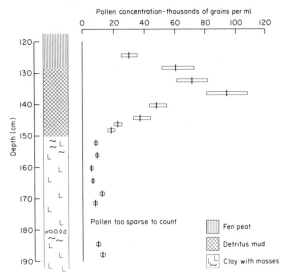

Fig. 6.3 The relationship between the total number of grains/ml and the type of deposit enclosing the pollen. Approximate 90% confidence limits for the counting error are shown for the absolute amounts. Data from Beanrig Moss, Roxburghshire.

number of grains/cm³ can be seen to vary between 6 000 and 93 000 from one sediment type to another. Obviously in this situation any change in the absolute density of a pollen type between two levels may be the result of the different sediment accumulation rate at the two levels. This is why most palynological work has been based on the relative frequencies of pollen types—so that a picture of the pollen changes can be obtained that is independent of the sediment enclosing the pollen.

Absolute pollen counts are useful if the sedimentation rate is constant or if the fluctuations in rate can be determined by radiocarbon dates closely spaced along the core. Pollen counts can then be expressed as numbers of grains of a species/cm²/year. This is termed *pollen influx*. A number of pollen diagrams using this method of presentation have now been published (e.g. Davis, 1967; Pennington and Bonny, 1970; Maher, 1972), and several different methods are available for the calculation of absolute pollen amounts (see Chapter 3).

Fig. 6.4 is a pollen diagram taken from Bonny

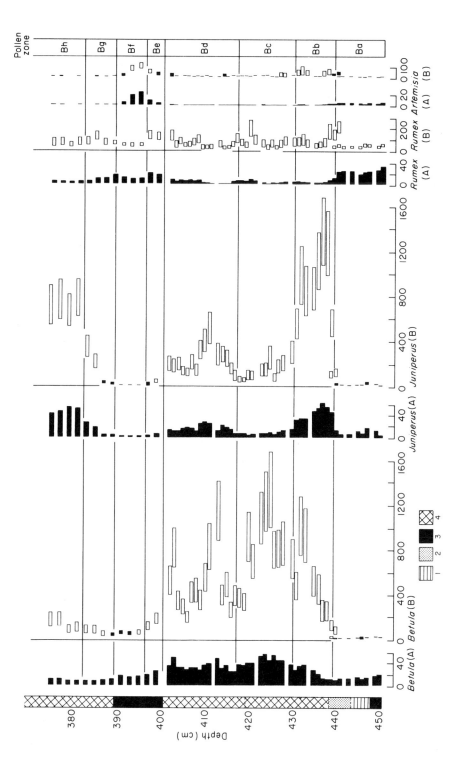

Fig. 6.4 Pollen diagram constructed for the comparison of absolute and percentage counts taken from Bonny (1972).
A = Percentage of total fossil pollen.
B = numbers of grains/cm²/year (pollen influx).
Approximate 90% confidence limits for the counting error are shown on the absolute amounts.
Stratigraphic symbols: 1. Organic mud; 2. Silt; 3. Clay; 4. Sapropel (algal) mud.

(1972) which compares percentage frequencies and absolute pollen amounts for a few selected taxa. The picture of the vegetational changes is much clearer when total pollen influx or influx for a particular species is available. At the base of the diagram, climatic change is registered by a large increase in pollen influx which is not evident from the pollen percentages. The 'swamping' effect of high *Betula* and *Juniperus* on *Rumex* and *Artemisia* percentages is also clear.

Despite their usefulness, absolute pollen counting methods have quite a few difficulties, and the essential set of closely spaced radio-carbon dates are at the moment so costly as to make the technique prohibitive. This means that percentages are still the most widely used method of expressing pollen data, and the many problems associated with interpreting them will now be considered.

Basically, the relative method of expression of pollen data represents the frequency of each pollen type as a proportion of a fixed sum. For example, each pollen type could be expressed as a percentage of the total pollen. However, this mode of expression has certain disadvantages. In many situations the pollen input from local flora fluctuates quite considerably as local changes, such as succession, take place. All proportional expressions suffer from the disadvantage that a change in one component will produce compensatory changes in all other components. Thus if local pollen proportions alter, this will result in changes taking place in the proportions of non-local types also.

Fig. 6.5 shows a hypothetical situation in which the total pollen assemblage is made up of three types, A, B and C. At the lowest depth these three types occur in equal proportions, each constituting 33% of the total pollen. Consider a situation in which the total pollen input of type A into the sediment were to vary while the total quantities of types B and C remained the same. If type A were reduced in its proportional representation by two-thirds, to 10%, the proportional frequency of B and C would rise by about 10% each. This compensatory movement of other pollen curves is inevitable where the total pollen input must be represented by 100%. Similarly, a real increase in A causes apparent decreases in B and C. Thus, in interpreting such a diagram one is presented with the problem of separating real changes from apparent ones.

An additional problem with proportional representation is that if the total pollen input changes, and if this is used as the basis of relative frequencies, then this change may not be apparent in the diagram, especially if proportions of different types remain unaltered.

Bearing these difficulties in mind, it is evident that the choice of pollen sum for use as a basis of representation is a critical one which may either facilitate or obscure the real changes underlying the fluctuating pollen curves. It is advantageous to separate the immediately local elements in the pollen assemblage and to eliminate these from the pollen sum, particularly where interest centres upon regional rather than local changes in vegetation. In most situations and during most of the post-glacial (Flandrian) period, excepting very recent times, the regional pollen is dominated by arboreal pollen. Because of this it has become conventional to count and to express pollen on an arboreal pollen basis. Obviously this cannot be done meaningfully when arboreal pollen is scarce, e.g. in tundra landscapes or

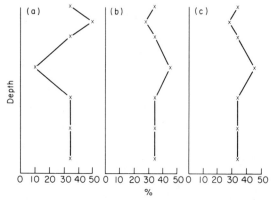

Fig. 6.5 Proportional representation of three pollen types, A, B and C, which together constitute a pollen assemblage. Expressions are in percentage total pollen.

where forest clearance has been very extensive. Where trees dominate the local pollen rain, e.g. *Alnus* or *Salix*, the local tree type can be removed from the pollen sum (Janssen, 1959). In this way it is possible to lessen the effects caused by alteration in the level of one component in the pollen sum. However, the basic disadvantage of percentage expressions still remains.

Where regional pollen is largely arboreal, it is convenient to express non-arboreal pollen types as a percentage of arboreal pollen, as in the case of the tree pollen types. However, complications arise when the regional pollen has a very small arboreal pollen fraction, for then local pollen types will be greatly exaggerated. For example, if a landscape becomes cleared of trees, the arboreal pollen component of the total pollen rain becomes smaller. Local pollen, however, remains at the same level. If local pollen is expressed as a percentage of arboreal pollen then it will rise as the arboreal pollen falls and this may give a false impression of vegetational changes.

An alternative is to express local pollen as a percentage of total pollen, or as a percentage of the arboreal plus local pollen sum.

Table 6.1 shows two hypothetical pollen counts taken at different layers in a situation where Ericaceae and Cyperaceae pollen types are local in origin and arboreal pollen is regional. In the time span between lower and upper samples

there has been regional clearance of woodland, as a result of which tree pollen is scarcer in the pollen rain. Although local pollen input has remained steady, one must now count far more local pollen grains in order to count 150 tree pollen grains. In the table the pollen types are expressed in three different ways, as described in the preceding section.

When local types are expressed on an arboreal pollen basis, their upper values are inflated to a considerable degree by their inverse relationship to the tree pollen. This effect is diminished when they are expressed either on a total pollen basis or as a percentage of the local + arboreal pollen sum. Usually the latter gives a preferable expression as it is less subject to fluctuations resulting from other sources of pollen (Wright and Patten, 1963).

Diagram construction

Several conventions are adopted when drawing pollen diagrams.

1. Vertical axis represents depth and horizontal axis the proportional abundance of the pollen types.
2. The stratigraphic sequence is represented diagrammatically in a column on the left-hand side of the diagram. This may prove of value in interpreting the pollen curves. Symbols for sediments do not, unfortunately, conform to

TABLE 6.1 *A hypothetical pollen count at two sampling levels (U = upper; L = lower) expressed in three ways.*

		Arboreal pollen	Shrub pollen	Ericaceae pollen	U/L ratio	Cyperaceae pollen	U/L ratio	Other
No. of pollen grains counted	U	150	75	200	10	50	5	25
	L	150	50	20		10		5
% Arboreal pollen	U	100	50	133	10·1	33	4·7	16
	L	100	33	13		7		3
% Total pollen	U	30	15	40	4·5	10	2·5	5
	L	64	21	9		4		2
% Arboreal pollen + local pollen	U	—	—	50	4·5	13	2·2	—
	L	—	—	11		6		—

any standard practice. Two schemes of representation which are frequently used are those of Godwin (1956) or Troels-Smith (1955).

3. The proportions of various pollen types is indicated at each level of sampling either by a bar histogram or by a point on a continuous curve. The latter method is the one which is perhaps most frequently encountered. Often the curves are made more distinctive by shading the area beneath them. Although general trends in pollen proportions may be more immediately evident from this type of diagram, it can be misleading. When two points are joined on the graph it implies that samples in intermediate levels would have proportions of pollen which would fall upon the line which has been drawn. This assumption is totally unjustified, hence it is preferable that known values should not be joined together in this way.

4. The horizontal scales of pollen proportions should be consistent throughout. This may be difficult since minor components may be overlooked. In such cases these components may be registered on exaggerated scales, but such exaggeration should be made conspicuous by omitting the shading beneath the curves, if such curves are employed (e.g. Birks, 1970, 1972a, 1972b).

5. Conventionally the pollen diagram is arranged in groups, with arboreal types first, followed by shrubs, then herbs and finally spores, etc. This method has the advantage that the approximate position on the diagram of any particular pollen type will be evident to the reader. On the other hand, there are instances where it is more sensible to group the pollen types either into local, regional and long distance transported groups, or into ecologically significant groups, e.g. open-habitat herbs, ruderals, aquatics, damp woodland types, etc. (Moore, 1973).

Some pollen diagrams have several pollen types superimposed upon one another in a single diagram (see Faegri and Iversen, 1964). This method has the advantage that much information can be conveyed in a minimum of space, but often it requires the sacrifice of clarity. Pollen diagrams where each type is displayed separately are far easier to understand than superimposed diagrams, even where distinctive symbols are used to identify the individual pollen curves.

6. It is useful to incorporate into a pollen diagram a summary chart which displays graphically the way in which the total pollen sum is divided into tree, shrub and herbaceous components at each level. This allows one to see at a glance any major changes in, for example, arboreal: non-arboreal pollen ratios, which may be of importance in the zonation and interpretation of a pollen diagram.

Statistical treatment of data

The percentage pollen diagram represents the observed proportion of each pollen type at successive levels. These observed proportions, as has been explained, are obtained by successive sampling and are estimates of the true proportions of each pollen type in the pollen rain at the time when the peat was forming. It is unlikely that the proportions of any given pollen type will remain constant at different depths (i.e. at different times in the past). Fluctuations may be due to two causes: (i) real changes in the composition of the pollen rain (which may in turn be caused by many variable factors) and (ii) observed fluctuations due to sampling errors.

Before one can begin upon a detailed interpretation of the pollen diagram, some attempt should be made to eliminate the fluctuations generated by the inadequacies of the sampling process and to reveal the true composition of the contemporary pollen rain. This requires a knowledge of the variation within the pollen data at any given sampling level and its comparison with the variation occurring between two levels.

The problems involved in the mathematical formulation of terms which will estimate these variations are considerable. Early attempts to obtain satisfactory statistical models are described by Faegri and Ottestad (1948), whose work forms the basis of the account given in

Faegri and Iversen (1964). More recent work by Mosimann (1962, 1963, 1965) has revealed certain flaws in the basic assumptions of Faegri and Ottestad. Mosimann (1965) provides revised formulae for the calculation of statistical confidence intervals for pollen data; a confidence interval is a measure of the reliability of a pollen proportion, its width being dependent upon the magnitude of sampling error at that level. Mosimann's method forms the basis of the account given here.

When one examines a population of pollen grains, one is faced with the necessity for identification, which can be represented as a choice, for example, is it *Quercus* or is it not *Quercus*? A choice with two alternatives of this type can be termed a *trial* and trials in which the outcome is uninfluenced by the outcome of any other trial, and in which the probability of a particular outcome is a constant, is termed a *Bernoulli trial*. In this context, *probability* means the likelihood of a particular outcome, e.g. if a population of 100 pollen grains contains 40 *Quercus* grains, then the probability p of the outcome of a trial being positive for *Quercus* is $40/100 = 0.4$. Conversely the probability of its not being *Quercus is* $60/100 = 0.6$.

If the total number of pollen grains counted remains constant (in this particular example, 100) then the probability distribution of the counts will be *binomial*. This means that if p is the probability of one trial outcome (*Quercus*) and q is the probability of the alternative outcome (non-*Quercus*) and if k trials (i.e. pollen grain identifications) are carried out, then $q + p = 1$ and the series is given by the expansion of $(q + p)^k$. In fact pollen sampling is not quite as simple as this because with each pollen grain one is faced with a number of possible outcomes. The question is not simply, *Quercus* or not *Quercus*?, but *Quercus*, or *Alnus*, or *Betula*, or *Pinus*, etc? Where there are several possible outcomes this is termed a *multiple Bernoulli trial*. If the total number of trials (i.e. pollen identifications) is kept constant, then each possible outcome (pollen type) obeys a binomial distribution individually. Jointly they follow a *multinomial distribution*.

The discussion so far assumes a fixed number of trials, i.e. a constant pollen sum to which each type contributes. In most pollen analyses this applies to the arboreal pollen (if a constant number of arboreal pollen grains is counted) but not to the non-arboreal pollen. If one considers a pollen type such as Gramineae, this is outside the pollen sum. In effect the grass grains are counted whilst a fixed number of 'marker' grains (usually arboreal pollen) are counted. This being so, a non-arboreal pollen type, such as Gramineae, follows a *negative binomial distribution*. If many pollen types are outside the fixed sum, which is usually the case, individual taxa follow a negative binominal distribution and jointly they follow a *negative multinomial distribution*.

It is necessary, as we have seen, to estimate the variation in data occurring within any given sampling level in order that this can be compared with between-level variation. Variation is expressed by the *variance* of the data and to obtain an estimate for this, replication is necessary. However, this does not necessarily mean that one has to count a large number of slides per level. Fortunately, a binomial variable with 'n' trials can be regarded as the sum of 'n' 1-trial binomial variables; in other words each trial can be regarded as a replicate. This being so, one can estimate within level variance from the pollen counts *within the sum* (i.e. those types which are binomially distributed). This variance can be compared with the variance of the true pollen proportion between levels.

The situation is similar when dealing with counts of types *outside the sum*, except that here the negative binomial variable with 'n' marker grains can be considered as the sum of 'n' independent negative binomial variables, each of which has a single marker grain. Replication is thus present in this type of count also. For each marker (e.g. arboreal) grain passed, a certain number of non-marker (e.g. Gramineae) grains will be passed, so each of the marker grains will provide a sample of non-markers. However, when considering the between-level variance of types outside the pollen sum, one must bear in mind that this variance will include the variance due to changes in the ratio of pollen type outside the sum to pollen within the sum. This ratio will vary from level to level.

The differences in the mathematical characteristics of pollen types inside and outside the sum means that they must be treated rather differently when analysed statistically.

Estimation of the true proportion and the confidence interval for a pollen type within the sum

At each particular level, one estimates the true proportion p of the pollen type within the sum by the ratio x/n, where x = number of grains of the pollen type and n = pollen sum. (For a discussion of the use of proportions, see Westenberg, 1967.)

If n is fairly large (the normal 150 count is adequate) then Mosimann (1965) has computed the confidence interval of the estimated proportion of the grain ($\hat{p}$), where

0·95 confidence limit

$$= \frac{\hat{p} + \left[\dfrac{(1·96)^2}{(2n)}\right] \pm (1·96)\sqrt{\begin{array}{c}[\hat{p}(1 - \hat{p})/n] + \\ [(1·96)^2/(4n^2)]\end{array}}}{1 + [(1·96)^2/n]}$$

0·99 confidence limit

$$= \frac{\hat{p} + \left[\dfrac{(2·58)^2}{(2n)}\right] \pm (2·58)\sqrt{\begin{array}{c}[\hat{p}(1 - \hat{p})/n] + \\ [(2·58)^2/(4n^2)]\end{array}}}{1 + [(2·58)^2/n]}$$

If 100 pollen grains are counted and 40 of them prove to be *Quercus*, the 95% confidence interval will be given by

$$\frac{0·4 + \left[\dfrac{3·84}{200}\right] \pm (1·96)\sqrt{\left[\dfrac{0·4(0·6)}{100}\right] + \left[\dfrac{3·84}{40\,000}\right]}}{1 + \left[\dfrac{3·84}{100}\right]}$$

$$= \frac{0·4 + 0·0192 \pm 0·0979}{1·0384}$$

$$= 0·4979 \text{ (upper limit)}$$
$$\text{or} = 0·3094 \text{ (lower limit)}$$

The 95% confidence interval of the pollen population is an important concept; it can be expressed in the following way.

In repeated sampling from a pollen population, 95% of the intervals calculated in this way will contain the true proportion of *Quercus* in the pollen sum. In other words, the interval 31%–50% in the example will contain the true proportion of *Quercus* 95 times out of a hundred, i.e. with a probability of 0·95. These intervals, then, are 95% confidence limits which we can place upon our observed value of 40% for *Quercus*, and between-level fluctuations with overlapping intervals can effectively be ignored (see Fig. 6.6).

If the 99% confidence limits are required then these can be calculated from the alternative formula. For this particular example, where 40% of a total of 100 grains are *Quercus*, the 99% confidence interval is:

Upper limit = 0·5286
Lower limit = 0·2839.

In other words, 99% of the observed proportions of *Quercus* in this particular pollen population will fall within the limits 28%–53%. If the estimation of the *Quercus* proportion in an adjacent level has a confidence interval which does not overlap with this range, then one can state with 99% probability that the assemblage does not belong to the original pollen population; in other words there is a real difference in the proportion of *Quercus* in the adjacent level.

Obviously, the higher the degree of precision required for this conclusion, the wider will be the calculated confidence interval, e.g. 99% intervals will be wider than 95% intervals. Intervals can be kept narrow only by increasing n, the number of grains within the pollen sum (see Fig. 6.7).

Confidence intervals can also be calculated for pollen types outside the sum, e.g. the non-arboreal types in a count based upon an arboreal pollen sum. Since these are distributed in a negative binomial fashion, the formulae necessary for the calculation will be different. In this situation one produces an estimated ratio of

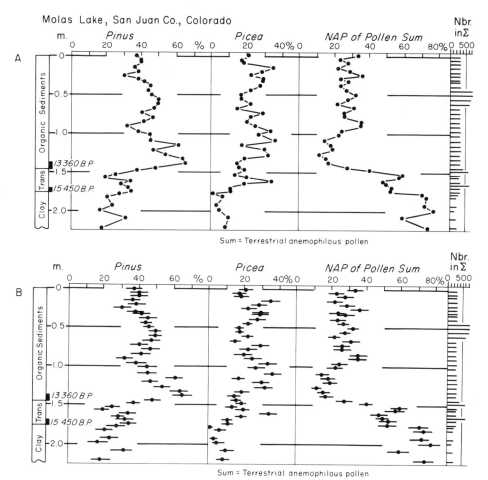

Fig. 6.6 Pollen diagrams from Molas Lake, Colorado. A. Conventional diagram; B. Diagram with 0·95 confidence intervals added. From Maher (1972).

the pollen type under examination to the total marker grains. This can be represented:

$\hat{u}$ (proportion to marker grains)

$$= \frac{x}{n} \quad \begin{array}{l}\text{(no. of type grains)} \\ \text{(no. of marker grains)}\end{array}$$

x is the number of grains of the type under investigation encountered whilst counting a total of n marker grains.

The 95% confidence interval is then given by:

$$\frac{\hat{u} + \left[\dfrac{(1{\cdot}96)^2}{(2n)}\right] \pm (1{\cdot}96)\sqrt{\begin{array}{l}[\hat{u}(1 + \hat{u})/n] + \\ [(1{\cdot}96)^2/(4n^2)]\end{array}}}{1 - [(1{\cdot}96)^2/n]}$$

and the 99% interval by:

$$\frac{\hat{u} + \left[\dfrac{(2{\cdot}58)^2}{(2n)}\right] \pm (2{\cdot}58)\sqrt{\begin{array}{l}[\hat{u}(1 + \hat{u})/n] + \\ [(2{\cdot}58)^2/(4n^2)]\end{array}}}{1 - [(2{\cdot}58)^2/n]}$$

For example, if 150 arboreal pollen grains are used as the basis of a count and during the course of this count 35 Gramineae pollen grains are found, it is possible to compute confidence limits for this proportion.

Here, $\hat{u} = \dfrac{x}{n} = \dfrac{35}{150} = 0{\cdot}233$

The 95% confidence interval is given by:

$$\frac{0{\cdot}2331 + 0{\cdot}0128 \pm 0{\cdot}0868}{1 - 0{\cdot}0256}$$

$$= 0{\cdot}3417 \text{ (upper limit)}$$
$$\text{or } = 0{\cdot}1635 \text{ (lower limit)}$$

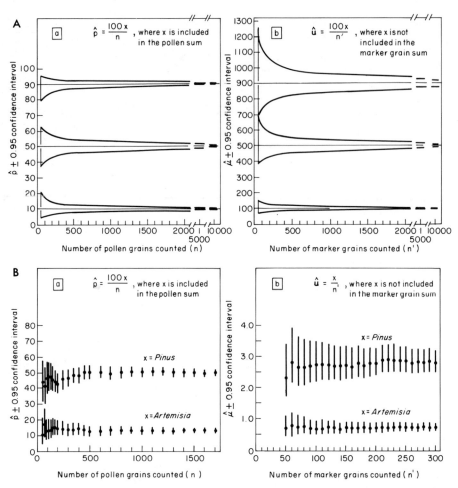

Fig. 6.7   Changes in the estimated proportion of a pollen type (p̂) and the 95% confidence interval of the proportion with an increasing number of grains counted A. theoretical; B. actual counts. a = pollen types included in the pollen sum. b = the same types when counted in relation to a marker grain (*Eucalyptus*). Based on equations by Mosimann (1965), after Maher (1972).

These are values for û, percentage expressions can be obtained by multiplying by 100.

Similar calculations using the formula for 99% limits give:

$$\text{Upper limit} = 0 \cdot 3876$$
$$\text{Lower limit} = 0 \cdot 1470.$$

These mean that when counting this population of pollen grains, the estimated value for Gramineae expressed as a percentage of arboreal pollen will fall between the limits 14·7% and 38·8% on 99% of the sampling occasions. This provides a means of comparing adjacent sampling levels with respect to this pollen type.

Although these calculations are somewhat involved, they are relatively simple to compute and are well within the capabilities of most programmable electronic calculators. The computation of these limits can also be achieved graphically by reference to the diagrams which have been published by Maher (1972).

On the pollen diagram these intervals can best be represented as a bar extending to the confidence limit on either side of the observed proportional value for the pollen type (see Fig. 6.6). In this way one can easily differentiate between random fluctuations in pollen curves and changes due to real alterations in the composition of the pollen assemblage. These real changes can then be used as the basis for constructing a zonation system for the diagram.

Pollen diagram zonation

It is conventional to divide the completed pollen diagram into a series of zones, each of which is considered to have some degree of internal uniformity. It must be recognized that any subdivision of the pollen diagram is an artificial activity, just as the subdivision of history into periods is artificial; however, it can be of considerable practical value since it facilitates the description and interpretation of changes in the pollen profile.

If we regard the zonation of a diagram in this way, then each diagram should be zoned upon its own characteristic features, without any recourse to the possible ecological or climatic implications of the zonation. Zones should be erected entirely upon the basis of the microfossils contained within them and should not be attempts at regional synthesis and correlation.

Pollen zones have not always been used in this way. For example, following his pioneer work in the palynological records of the British Isles, Godwin (1940) erected a zonation system which he considered applicable to the whole of southern Britain. This system was based upon the changing proportions of tree pollen in the diagram; Fig. 1.1 (page 4) gives a summary of the major features upon which zone boundaries were based. The zonation system applied only to the closing phases of the last (Devensian) glaciation and the post-glacial (Flandrian) period. The zones were labelled I–VIII from the base. Similar systems were erected on the continent of Europe (Firbas, 1949) and in Ireland (Mitchell, 1942; 1956).

Although the Godwin zonation system was based upon fluctuating proportions of tree pollen, it soon came to be associated intricately with the climatic changes underlying these fluctua-

tions. For example, Godwin's zone VIIa became identified with the Atlantic period of Blytt and Sernander and thus had immediate climatic connotations. In turn the zones became associated with radiocarbon datings, despite the lack of evidence for synchroneity of the zone horizons in different parts of the country.

The pollen zone thus came to be regarded as a period of time, having a particular type of climate resulting in certain features recorded in the proportion of various tree pollen types. Godwin himself was aware of the difficulties resulting from this concept of the pollen zone. When describing pollen data from parts of the country which were climatically distinct from that of southern Britain he was careful to use local zonation systems which might be correlated only tentatively with the basic scheme. An example of this approach is provided by his work on Tregaron Bog in west Wales (Godwin and Mitchell, 1938).

Work by Iversen (1941) and later by Turner (1962) made it evident that the pollen changes which were used to define the more recent zone boundaries (e.g. the elm decline at the zone VIIa/VIIb transition and the fall in lime at the VIIb/VIII boundary) were not climatically determined, but resulted from the activity of man. These findings resulted in the undermining of the temporal connotations of the more recent pollen zones, for, although climatic changes may have been widespread and synchronous, it was unlikely that anthropogenic effects upon vegetation should be more than local in extent and it was probable that they had occurred at different times in different places. As a result, palynological work dealing with changes in the past 5000 years abandoned the classic zonation system in favour of local zonation schemes which were regarded as valid only within the area of study (see, for example, Conway, 1954).

Often the earlier zones of the Godwin system were also difficult to recognize, especially in regions outside southern Britain, and local zonation systems have been produced, e.g. for the Cumberland lowlands (Walker, 1966). Never-

theless the vast majority of published pollen diagrams from the British Isles have adopted the
Godwin zonation system either in a pure or a
modified form. This has served to simplify the
general picture of vegetational change, but often
the simplification has proceeded to the point of
being misleading. In effect the zonation of pollen
diagrams has become an exercise in fitting the
data from the site under investigation into a pre-
determined mould. Undoubtedly the significance
of many local and regional variations in the general
picture has been lost as a result of this concern for
conformity on the part of British palynologists.

In North America, the size of the country and
its diversity of climate and biota led people to
expect a variation in the developmental history
of vegetation in different localities. As a result
the concept of a pollen zone as a biological/
ecological/climatic/temporal division in the history of the flora was of little practical value. They
developed instead a concept of a zone which is
defined entirely upon its internal microfossil constituents and which need have no existence outside the region in which it is first recognized and
defined. The concept was finally formulated by
E. J. Cushing who termed the division so created
an '*assemblage zone*'. He subsequently demonstrated the use of his formulation when considering the pollen stratigraphy of sites in Minnesota
dating from the closing stages of the final
(Wisconsin) glaciation (Cushing, 1967b). The
concept has recently been introduced into British
palynology by H. H. Birks (1970; 1972a; 1972b)
where it has proved of considerable value in
elucidating the regional variations in the
Flandrian history of vegetation in Scotland.

The use of this system means that one can zone
each diagram upon its own features without
reference to any other pollen diagrams. Since
some pollen types are locally derived it is possible
that a pollen assemblage zone will simply refer to
the state of the vegetation in the immediate vicinity. This may in itself be of value, e.g. when one
is concerned with elucidating local successional
or retrogressive developments in vegetation (e.g.
Iversen, 1964; Moore, 1973). More frequently

one is interested in regional changes in vegetation, in which case pollen types which are known
to be local, or which have local components, can
be ignored when constructing the zonation
system. The zones resulting from this treatment
can be regarded as regional pollen assemblage
zones, and their counterparts may be expected in
other diagrams from the area. Usually these assemblage zones are named after the most conspicuous pollen or spore components rather than
being numbered (see Birks, 1970).

One advantage of this type of zonation system
is that additional zones and 'sub-zones' can be
constructed without disrupting a numbered
sequence of zones. The system is more acceptable
ecologically than the traditional system because
it is based entirely upon the pollen curves, which
themselves reflect the surrounding plant communities. It is flexible and it allows for local variations, which are to be expected in conditions of
diverse topography, soil and microclimate.

Figs. 6.8–6.10 show a pollen diagram of
Flandrian deposits from west Wales (Moore,
1972). This has been zoned upon its inherent
features and the zones labelled Ga to Gh, start-
ing from the base (the G stands for Gwarllyn, the
name of the site). Some of these zones can be
equated roughly with the Godwin zonation
system, though this does not necessarily imply
synchroneity of zone boundaries; rather it
indicates a similar sequence of vegetation to that
frequently found in southern Britain. Some of
the earlier zones are defined by criteria quite different from those used in the Godwin system,
hence these have no equivalents.

The pollen assemblage zones at Gwarllyn are
as follows:

> *Zone Ga*   Grass/sedge/juniper zone
> *Zone Gb*   Birch/grass zone
> *Zone Gc*   Birch/pine zone
> *Zone Gd*   Birch/pine/hazel zone
> *Zone Ge*   Oak/hazel forest zone
> *Zone Gf*   Oak/hazel/birch forest zone
> *Zone Gg*   Oak/hazel/alder forest zone
> *Zone Gh*   Forest/ruderal zone

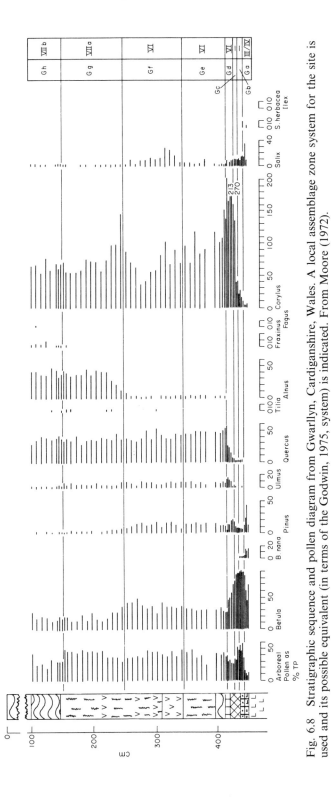

Fig. 6.8 Stratigraphic sequence and pollen diagram from Gwarllyn, Cardiganshire, Wales. A local assemblage zone system for the site is used and its possible equivalent (in terms of the Godwin, 1975, system) is indicated. From Moore (1972).

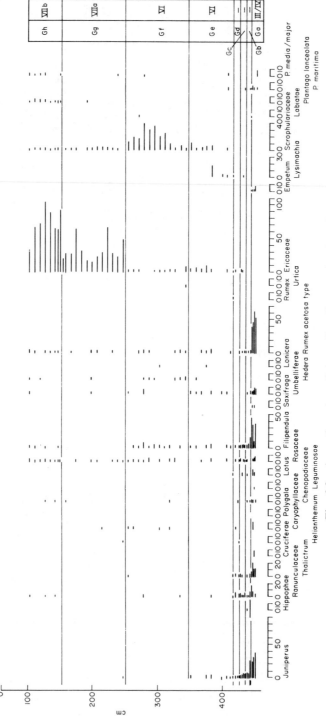

Fig. 6.9 A continuation of the pollen diagram for Gwarllyn.

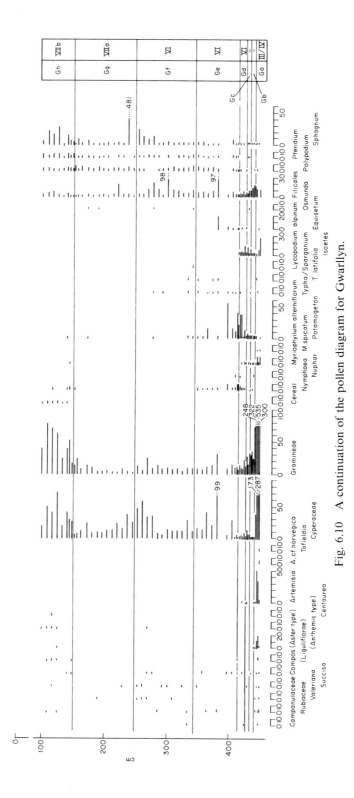

Fig. 6.10   A continuation of the pollen diagram for Gwarllyn.

There is a hiatus in the stratigraphy at the Gg/Gh boundary, peat having been cut back and regenerated. There is therefore a temporal gap of unknown duration between the two deposits.

The early zones in this diagram must be regarded as transitional. They almost certainly represent a vegetation in a state of dynamic flux as new species immigrated and expanded their populations and range under the influence of rapidly changing climate. Nevertheless, each zone is distinct from all others in terms of the assemblage of pollen found within it.

These early zones (Ga–Gc) cannot be equated strictly with any of the Godwin zones on the basis of their pollen content. The behaviour of certain pollen types, particularly *Corylus*, is quite unlike the pattern exhibited in most other sites in southern Britain (Moore, 1972). This almost certainly reflects the microclimatic pecularities of this site in western Britain with its strong maritime influences. Obviously it would have been misleading to attempt to zone this diagram on the basis of Godwin's criteria.

Regional variations in pollen diagrams which make difficult the application of the traditional zonation system are even more marked in Scotland. Here the use of the pollen assemblage zone concept is vital. An example of this is given in Fig. 6.11, which is part of a pollen diagram from Abernethy Forest in the Cairngorm area of Inverness-shire (H. H. Birks, 1970). Radiocarbon dates indicate that the diagram covers the entire Flandrian period. Comparison with the generalized diagram for southern Britain (Fig. 1.1, page 4) shows that the entire pattern of pollen type frequencies is totally different at this site, hence it would not be merely inadvisable to attempt to apply the Godwin zonation system, it would be impossible. Hilary Birks has therefore applied a local zonation system which she has described in terms of pollen assemblage zones. The contrast between this site and those in the south becomes even more apparent when it is realized that the *Pinus* assemblage zone began at about 5000 BC, i.e. soon after the zone VI/VIIa boundary in England (e.g. Scaleby Moss, Cumberland).

One must conclude, therefore, that at least the initial zonation of a diagram should be undertaken without reference to other pollen diagrams, but should be based upon the inherent features of the data in question. Subsequently, a comparison and tentative correlation with other diagrams can be exceedingly informative and valuable.

Objective zonation

Pollen diagram zonation is essentially subjective and is performed solely for the sake of convenience in description and interpretation. The pollen diagram represents a set of continuous variables which must be broken at selected points. Since the zones are, by definition, relatively uniform in their pollen constituents, the zone boundaries must of necessity pass through those portions of the pollen curves where change is most marked. Usually the boundary is placed where there are several concurrent changes in a number of pollen types.

This requirement for the position of the pollen zone boundary has provided the opportunity for the division of pollen diagrams into zones by the use of multivariate statistical analysis, usually with the aid of computers. In essence there is no reason why each level of sampling should not be regarded an 'individual' possessing certain 'attributes', i.e. pollen characteristics. The 'individuals' can then be classified on the basis of their 'attributes' in an objective manner by the use of statistical classificatory systems (e.g. Dale and Walker, 1970). This use of multivariate, classificatory techniques is precisely equivalent to their use in phytosociology where quadrats are classified into groups on the basis of the species they contain, or vice-versa. There is one major difference, however, between a quadrat and a pollen sampling level; in the case of the former there are no restrictions upon the ways in which they can be grouped to provide meaningful ecological information. In the case of the latter, however, the samples occur in a linear sequence which is determined by stratigraphy, as a result

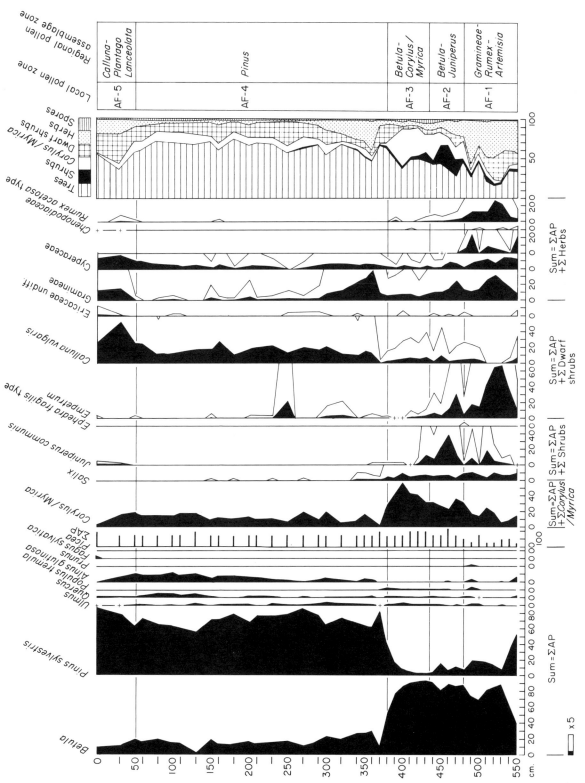

Fig. 6.11 Pollen diagram from Abernethy Forest, Spey Valley, Inverness-shire. AP = Arboreal pollen. From H. H. Birks (1970).

of which it is preferable that the samples should be grouped in such a way that consecutive samples are either grouped together or occur in adjacent groups (or 'zones').

Gordon and Birks (1972) have overcome the problem by introducing a 'stratigraphical constraint' into their computer program which respects the linear sequence of pollen samples. They have used a number of different classificatory techniques with a considerable degree of success. In this way the subjectivity involved in selecting zone boundaries can be eliminated.

It should be added, however, that for most practical purposes such elaborate techniques are not essential. Indeed, the objective techniques of Gordon and Birks have usually served to confirm the zonation systems derived from subjective methods (see Chapter 8). The validity of accurate pollen data is not affected by misplaced zone boundaries, though their implications may be obscured.

Each level of sampling in a profile contains pollen which can be resolved into a 'spectrum' of different pollen types. The pollen diagram is a collation of these spectra which gives a temporal dimension to what would otherwise be a static picture. We interpret the pollen diagram, therefore, as a dynamic record of vegetational history.

Underlying this type of interpretation is the basic assumption that the pollen spectrum at a particular level in some way represents the contemporary, pollen-producing flora which was contributing to the pollen rain. The process of interpretation often proceeds one step further and attempts to reconstruct the environment, both physical and biotic, in which that vegetation type grew. This second stage of interpretation demands a further assumption, namely that vegetation in some way reflects its environment.

Both of these assumptions have certain limitations and these limitations will provide the basic theme of this chapter.

The first assumption

The first assumption underlying the interpretation of pollen diagrams is that the pollen provides some indication of the vegetation existing in an area. There are many factors which influence the relative abundance of a particular pollen type at a given level and which therefore influence the precise relationship between vegetation and pollen rain. These factors will first be listed and then considered individually.

1. The proportion of other pollen types in the spectrum.
2. The land area from which the pollen is derived.
3. The dispersal efficiency of the pollen type.

4. The pollen productivity of the parent species.
5. The local environment of the parent plant.
6. The time of flowering of the parent plant.
7. The morphological and physiological condition of the parent plant.
8. The density of the parent species in the local and regional environment.

Proportional representation

Proportional representation not only presents a problem in the expression of pollen data, it also confuses their interpretation. It was explained in Chapter 6 that the percentage abundance of a pollen type will be influenced by the abundance of other types (see, for example, Fig. 6.2, page 80). For a pollen type counted inside the sum (e.g. a tree pollen grain in an arboreal pollen count) the level of other types included in the sum will influence its level. For a pollen type counted outside the basic sum it is the abundance of that particular type in relation to the types composing the basic sum that is being expressed. Thus absolute changes in pollen input of any type within the sum will affect the levels of other components of the sum and all the components outside the sum that are expressed on the same basis.

Faegri and Iversen (1964) have also pointed out that the percentage representation leads to an uneven expression of any absolute changes in the abundance of a pollen type, depending upon the proportion of that type already present. This can be illustrated by a hypothetical example. Consider a site which receives only two pollen types, A and B, and which receives 90 grains of species A and 10 grains of species B per unit area per unit time. The representation would obviously be 90% A and 10% B. If the pollen input of species

A doubled because of some vegetational change, then the situation would be:

A   180 grains/area/time = 94·7%
B   10 grains/area/time = 5·3%

A doubling of the pollen input of A has raised its representation by only 4·7%. If, on the other hand, B were to double its pollen input the representation would be:

A   90 grains/area/time = 81·8%
B   20 grains/area/time = 18·2%

The second situation has probably involved far less change in the contemporary vegetation than the first, but it has had a greater influence upon the proportional representation of the two types. Faegri and Iversen refer to this statistical effect as an illustration of the 'law of diminishing returns' for as a species approaches 100% representation a large absolute change in pollen input will cause very little change in its representation. The same principle operates in reverse when a species has a very low representation: a halving of its pollen input will cause only small changes in representation. Halving the input of a species with a high proportional representation has marked effects.

A   90 grains/area/time = 94·7%
B   5 grains/area/time = 5·3%

A   45 grains/area/time = 81·8%
B   10 grains/area/time = 18·2%

Comparison of these figures with the original ones shows that the effect of halving type A upon the proportional representation of the two is precisely the same as doubling the input of type B.

These effects mean that one can never interpret the level of a particular pollen type in isolation if it is expressed in proportional terms. If one were to adopt a totally cynical attitude it means that one cannot interpret the level of any pollen type with any degree of certainty. To some extent this is true, but a careful study of the relative fluctuations of all the influential pollen types can assist in the separation of 'real' from statistical changes. If cynicism leads to a more cautious approach to the interpretation of pollen diagrams, then it should be encouraged.

The grave difficulties involved in the interpretation of diagrams expressed in proportional terms has led to an increasing emphasis upon absolute pollen data, where the precise numbers of pollen grains falling per unit area per unit time can be calculated. As explained in Chapter 6 this requires a knowledge of sedimentation rate together with that of the pollen density within the sediment. However, if such data are available then this initial problem in pollen diagram interpretation can be overcome.

The origins of fossil pollen

Before one can begin to interpret the fossil pollen assemblage in terms of the surrounding vegetation, one must consider where these grains originated and how far they have travelled. The essential question is how great an area of vegetation is represented within the assemblage.

Functionally, the pollen grain represents genetic information which needs to be transferred from the male reproductive structure to the female organ of the same or a different plant of the same species. Once such pollination has been effected, the germination of the pollen grain and subsequent fertilization can take place. Many complex mechanisms have evolved which aid this process of pollination (see Proctor and Yeo, 1973) and it will be necessary to consider some of these in a later section when the quantity of pollen produced by individual plants will be discussed. Whatever mechanism is employed, it must ensure the transport of pollen at least as far as the next individual of the species (for outbreeding species).

In practice it is often difficult to tell how far viable pollen can be transported by a vector such as wind, and still effect pollination and fertilization. Occasionally, however, the study of genetics and taxonomy of plants reveals the spatial distances over which the flow of genetic material can take place by means of pollen transport. The genus *Pinus* on the Pacific coast of North America provides a good example (Mason, 1949). *Pinus remorata* is a species endemic to the islands of Santa Cruz and Santa Rosa, over thirty miles off the coast of North America. On the mainland, a closely related species, *P. muricata* is found, and in late Pleistocene times this species has invaded Santa Cruz and Santa Rosa with the result that hybridization has occurred. There is evidence which suggests that there has been some genetic infiltration of *Pinus remorata* traits into the mainland population of *P. muricata*, but since *P. remorata* has never succeeded in crossing to the mainland one suspects that this gene flow is due to wind-borne pollen arriving at the mainland either from *P. remorata* or from the hybrids. This pollen is evidently capable of effecting pollination and fertilization in the mainland populations of *P. muricata* 30 miles away.

Here, then, is a situation in which wind pollination is employed and where the capacity for viable pollen transport over a distance of more than thirty miles is evidently proving worthwhile in terms of the conservation of particular genetic combinations. This being so, it is understandable that the pollen grains of species utilizing wind pollination mechanisms should have evolved structural modifications which aid their transport over long distances.

Some studies have been made on the patterns of dispersal and deposition of wind transported pollen grains (e.g. Wright, 1953). Fig. 7.1 shows the type of curve obtained when the proportion of pollen deposited is plotted against distance from the source. This type of distribution, where there is a high likelihood of individuals falling close to the source, is termed *leptokurtic*.

Different species of plant differ in the precise form of their curves, but these differences can usually be described by reference to their standard deviations, as in the case of the similar normal distribution. Wright (1953) has calculated standard deviations for the leptokurtic pollen dispersal curves of a number of wind pollinated tree species. For example, *Pseudotsuga menziesii*

(Douglas fir) has a standard deviation of 14 m, *Picea abies* (Norway spruce) 43 m and *Ulmus americana* (Elm) over 300 m. However, these figures are of limited value to the palynologist except in making comparisons of the efficiency of dispersal in different species. Such figures will vary with atmospheric conditions and according to whether samples are taken upwind or downwind of the source. The topographic situation of the source will also influence dispersal patterns (see page 111).

The most important initial problem for the palynologist faced with a fossil pollen assemblage is that of understanding where the pollen has originated.

Tauber (1965) has constructed a model accounting for the various mechanisms whereby pollen may arrive at a mire or a lake surface (see Fig. 7.2). In this model he considers a site surrounded by forest (as were many sites for the bulk of the post-glacial period). Pollen falling out on the site consists of three major components:

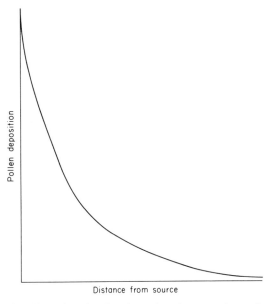

Fig. 7.1 Graph showing the decrease in pollen deposition as one moves away from the source of an anemophilous pollen type. The distribution is leptokurtic.

1. Trunk space component. This is the pollen which falls vertically from the tree canopy or is produced by shrubs and herbs beneath the canopy. Some of this pollen may be transferred above the canopy by gusts of wind in clearings, but most of it is deposited beneath the canopy. Where the forest is interrupted by a mire or lake site the air currents which pass through the trunk space will deposit some of their pollen load on to the surface.

   Air currents beneath the canopy are greater than those within the canopy once leaf emergence has taken place (Geiger, 1965; Fig. 7.5), but are considerably lower than those above the canopy. This slow-moving air is thus less able to carry pollen than the faster-moving air above, hence deposition occurs in relatively short distances.

2. Canopy component. Some of the pollen produced within the canopy, or escaping from below, will be carried along by air currents above the canopy itself. A proportion of this component may be transferred by thermals to high altitudes where it will be carried along by high altitude winds. Some may be trapped by eddies in the surface of the canopy and slowed

down to such an extent that it sinks through the canopy and joins the trunk-space component.

In Tauber's opinion, this component may be carried along in such a way that it is taken over the top of any small mire or lake and very little of the pollen load is liable to settle out on its surface (see Fig. 7.2).

3. Rain component. Pollen, along with other dust particles, may act as nuclei around which water droplets form. When such droplets descend as rain they collect more dust and pollen on the way. This mechanism of pollen fall-out probably accounts for the main transfer of pollen from higher altitudes on to the mire surface.

   In order to complete this model of Tauber's, we could add two further sources of pollen into the fossil assemblage.

4. Local component. Pollen from aquatic plants growing in a lake, or from semi-aquatic species growing on the surface of a mire may be expected to deliver a large proportion of their pollen production to the surface surrounding them (see Fig. 7.1). In certain situations this component can constitute a large proportion

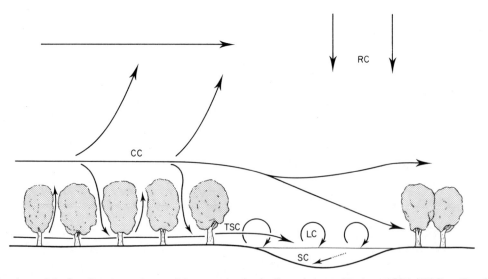

Fig. 7.2 A model of pollen input into a lake or mire basin (based upon Tauber, 1965). TSC = Trunk space component; CC = Canopy component; RC = Rain component; LC = Local component; SC = Secondary component.

of the total pollen input, e.g. in eutrophic lakes where macrophytes are abundant and on blanket mires where such genera as *Eriophorum*, *Calluna*, *Erica* and *Molinia* may dominate the local flora. Often this component can be identified by reference to the macrofossil content of the peat (see Chapter 6).

5. Secondary component. Where a surface receives drainage water from a surrounding catchment, it is liable to receive pollen which has been deposited elsewhere and subsequently mobilized and transported. This presents two problems. If the transported pollen consists of recently deposited grains which have never been incorporated into a sediment, then its addition to the fossil assemblage effectively increases the non-local component of the assemblage. This need not affect the interpretation in any substantial way unless the secondary input of pollen is very large in comparison with the primary input to the surface and if its secondary origin is not recognized.

A further problem arises if the secondary pollen has been incorporated into a sediment within the catchment and has subsequently been eroded and transported in surface water. The secondary deposition of these grains may cause considerable confusion because they will be of greater antiquity and will possibly have been derived from quite different vegetation types. In Chapter 2 this problem was discussed and the possibility of using the state of preservation of such grains or their fluorescence properties as indicators of their secondary nature was put forward.

Any fossil assemblage may contain any or all of these five components and the area represented within an assemblage will depend upon the importance of each of the components in the process of pollen deposition at the site. This, in turn, depends upon such factors as the size and shape of the mire or lake surface, the configuration of the catchment and the type of vegetation surrounding the site. In particular, whether a site is surrounded by forest or by short

vegetation will obviously affect the relative importance of trunk space and canopy components and thereby will influence the proportion of the rain component falling on the surface.

To obtain some estimate of the relative importance of these various components at a given site, a study of contemporary pollen rain is necessary. Two main approaches have been used for this type of study, either the analysis of superficial peat or lake sediment samples or the trapping of air-borne pollen and spores in specially contrived pollen traps. The former method is simpler from the point of view of collection of samples, but it does depend upon the availability of suitable lake or mire sites. Where such sites are not immediately available, moss polsters have been used with a certain degree of success. However, if such polsters have developed slowly in aerobic conditions then it is possible that differential decay processes have been at work and that the remaining fossil assemblage will not be representative of the original pollen rain.

Surface samples represent the accumulation of pollen input over one, or several, seasons. Spore traps, on the other hand, need to be used for at least a year before the data obtained from them can be used to interpret fossil assemblages. The aerodynamic properties of such traps must also be taken into consideration.

A number of different types of trap have been derived for sampling the pollen and spores carried by air currents, several of which have been discussed by Lewis and Ogden (1965). Some samplers act simply as rain collectors, but because of their aerodynamic properties they are unlikely to sample adequately the particles suspended in the air. The collection efficiency of such traps varies with meteorological conditions, particularly wind (Ranson and Leopold, 1962). Davis (1967) has used similar, open-necked traps for the determination of pollen deposition within lakes. She found that the number of grains accumulating in these traps was linearly correlated with the area of aperture, suggesting that the efficiency of the trap is not altered if different-sized traps are used.

Tauber (1967a) modified this trap type for use in turbulent airflow conditions. The modification basically involves the use of a curved lid of greater diameter than the top of the vessel, in which a hole of slightly smaller diameter is bored. This trap has also been used in aquatic conditions and its efficiency in such situations has been tested by Peck (1972). She found that the efficiency of capture varied with the size and weight of the pollen grains involved and with the velocity of flow. Smaller grains were trapped less efficiently than larger ones under low flow conditions, but differences became less as the water flow rate increased. The simplicity of construction of this type of trap, together with its robustness and inconspicuous nature, have made it a popular and widely used instrument for measuring pollen deposition over relatively long periods, i.e. for periods greater than a day.

Short period and continuous records of pollen deposition, e.g. for monitoring variation in pollen concentration during a day, often utilize the Hirst spore trap (Hirst, 1952) or some similar instrument, which is based upon the movement of a slide across an orifice through which air is blown. Pollen is impacted upon the slide and its variation with time of day can be calculated.

A final type of sampler involves the filtering of air sucked in at a known rate. The problem with this type of trap is that the intake rate is affected by windspeed and the direction in which the intake orifice is facing.

These various types of spore sampler are obviously designed for various different purposes. The palaeoecologist is concerned mainly with long-term accumulation of pollen from a particular vegetation type rather than the diurnal or even seasonal variation in pollen concentration. Because of this, most modern pollen deposition studies which have been undertaken with palaeoecological interpretation in mind have used either moss polsters (e.g. Hansen, 1949; Rymer, 1973; Birks, 1973a) or some form of rain gauge, open-necked collector or Tauber trap (see Wright, 1967).

Extensive surveys of contemporary pollen spectra have been undertaken in many parts of North America and Europe. Some of the more important surveys will be listed here but they cannot all be considered in detail.

Canada: Ritchie and Lichti-Federovich (1963; 1967), Lichti-Federovich and Ritchie (1965; 1968), Bartley (1967), Terasmae (1967)
USA Minnesota: Janssen (1966; 1967a; 1967b)
　　Michigan: Davis, Brubaker and Beiswenger (1971), Janssen (1973)
　　Washington: Heusser (1969)
　　Wyoming: McAndrews and Wright (1969)
　　North Carolina: Whitehead and Tan (1969)
Iran: Wright *et al.* (1967)
India: Singh, Chopra and Singh (1973)
S.E. Asia: Flenley (1973)
New Zealand: Moar (1970)
Iceland: Rymer (1973)
Wales: Hyde (1952)
Scotland: Birks (1973a; 1973b), Turner (1964b), O'Sullivan (1973)
Scandinavia: Berglund (1973)

These surveys differ from one another in many respects. Some of them cover wide areas of ground and attempt to compare current pollen spectra in different vegetational zones, e.g. prairie, boreal forest, tundra, etc. Others are concerned with variations in pollen sedimentation within a single mire in order to determine how local conditions affect the pollen spectrum. It may be useful to consider one of each of these types of survey in more detail so that their respective relevance to pollen diagram interpretation can be evaluated.

One of the most extensive studies from the geographical point of view has been that of Lichti-Federovich and Ritchie in Canada. They have studied both airborne pollen and superficial samples from lakes over a considerable area of central Canada. The data presented in Table 7.1 is abstracted from a survey of samples from over 100 lakes which was published in 1968. Fig. 7.3 shows the localities of the selected sites, together with the boundaries of the major vegetational zones. Table 7.1 gives a brief selection of analyses

TABLE 7.1   *Pollen spectra from selected sites in Western Interior Canada, shown in Fig. 7.3. From Lichti-Federovich and Ritchie (1968). Figures are percentages of land pollen.*

Zone Station	T 1	FTU 15	FTL 23	OCF 25	CCF 40	MF 70	DF 105	AP 96	G 109
Picea	5·6	30·6	33·2	34·2	30·0	12·4	0·3	1·7	1·2
Pinus	14·0	20·4	26·2	25·6	58·7	21·4	6·1	10·4	14·8
Betula	42·6	16·0	5·0	14·6	5·5	11·2	23·5	1·1	1·0
Alnus	7·0	15·0	6·4	12·2	1·6	9·4	3·0	2·0	0·8
Salix	1·4	0·1	0·8	1·2	0·4	2·9	4·0	3·3	1·4
Populus	—	0·2	0·2	—	—	3·3	4·0	2·4	2·8
Carya	—	0·2	—	—	—	—	—	—	—
Myrica	—	0·2	—	—	—	0·3	—	—	0·1
Fraxinus	—	—	—	—	—	0·1	2·0	0·5	5·5
Carpinus/Ostrya	—	—	—	—	—	—	—	—	—
Juniperus/Thuja	—	—	—	—	—	0·2	—	0·3	0·1
Quercus	—	—	—	—	—	0·4	2·3	0·1	0·1
Ulmus	—	—	—	—	—	0·4	0·5	0·3	0·7
Acer	—	—	—	—	—	0·2	—	0·2	0·3
Corylus	—	—	—	—	—	0·3	6·6	0·1	—
Symphoricarpos	—	—	—	—	—	—	0·3	0·1	—
Shepherdia	—	—	—	—	—	—	—	0·1	—
Juglans	—	—	—	—	—	—	—	—	0·1
Total tree	62·2	67·4	64·6	74·4	94·2	49·4	38·6	16·7	26·5
Total shrub	8·4	16·2	7·2	13·4	2·0	13·1	13·9	5·9	2·4
Gramineae	2·6	0·6	1·4	1·0	1·2	15·4	7·3	17·7	13·9
Chenopodiaceae	1·2	0·4	0·6	0·8	0·7	2·0	6·6	13·3	18·6
Ambrosieae	2·0	1·2	2·6	0·2	0·1	0·1	5·1	3·4	2·6
Artemisia	1·4	1·4	2·0	1·4	0·8	2·2	24·7	20·8	24·3

which were made of surface lake sediments through these zones; only one sample is quoted here for each zone. These figures can be compared with those of Table 7.2 which relate to the land surface area by different vegetation types or genera as determined from aerial photographs.

Although the two tables show some broad correspondence to one another it is their points of disagreement which are perhaps more striking. From this data alone it becomes obvious that the interpretation of a pollen spectrum must of necessity be a complex business. There are many reasons why discrepancies could occur between regional vegetational cover and the spectrum of pollen fallout and one of these is the point under consideration in this section, namely the location of origin of the pollen grains.

The broad correspondence of Tables 7.1 and 7.2 referred to above can be taken as an

TABLE 7.2   *Composition of the vegetational zones of Fig. 7.3, mapped from aerial photography. Figures are percentage land area. From Lichti-Federovich and Ritchie (1968).*

Zone	FTU	FTL	OCF	CCF	MF	DF
Heath tundra	53·2	30·2	8·8	—	—	—
Sedge	3·3	42·2	1·4	2·1	3·1	—
Picea	41·2	23·7	80·0	31·9	37·5	—
Alder/willow	—	—	6·1	3·4	0·8	6·4
Pinus	—	—	1·5	33·1	1·4	—
Betula	—	—	0·2	2·4	3·5	11·7
Larix	2·3	3·9	2·0	0·2	0·6	—
Abies	—	—	—	7·1	1·2	—
Populus	—	—	—	19·8	51·7	38·2
Quercus	—	—	—	—	0·1	15·3
Ulmus	—	—	—	—	—	8·9
Fraxinus	—	—	—	—	—	13·8
Grass–herb	—	—	—	—	0·1	5·7

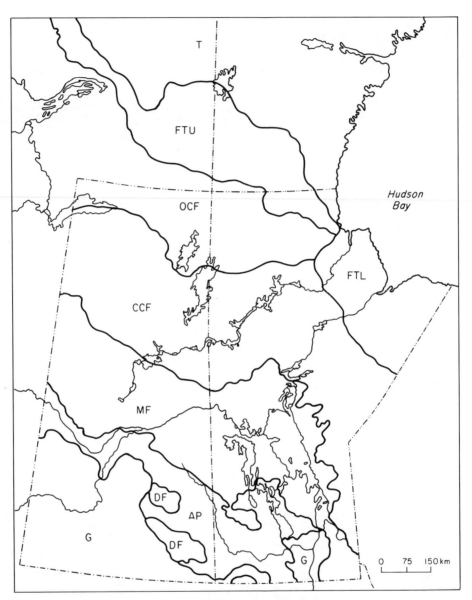

Fig. 7.3   Map of Western Interior of Canada showing selected sampling sites and the main vegetation zones (modified from Lichti-Federovich and Ritchie, 1968). T = Tundra; FTU = Upland forest tundra; FTL = Lowland forest tundra; OCF = Open coniferous forest; CCF = Closed coniferous forest; MF = Mixed forest; DF = Deciduous forest; AP = Aspen parkland; G = Grassland.

indication that regional pollen is of considerable influence in many of these spectra. Some of the points of difference may be attributable to the long-distance movement of pollen between vegetational zones, which presumably arrives at the point of sampling as a result of washout from the atmosphere.

Pinus is a good example of a taxon which is not detectable on aerial photographs from the forest tundra zone, yet the pollen spectrum of this region contains over 20% *Pinus* pollen. *Pinus* even maintains a level of 14% in the treeless tundra area and *Picea* here is still 5·6%. *Betula* is not as easy to interpret because only tree birches were included in the aerial survey. Dwarf birches occur in the tundra but would have been mapped under the general heading 'heath-tundra'. *Larix* behaves in the opposite way, since it occupies a detectable proportion of the land surface in the forest tundra zone and yet is not represented in the pollen spectrum. This, however, is probably to be explained by the poor preservation of *Larix* pollen.

Long-distance transport of the pollen of deciduous trees into boreal and tundra regions does not appear to be very significant, however the deciduous forest region still receives an input of *Picea* and *Pinus*. In this region, disturbance of the natural vegetation by man could account for such occurrences.

Discrepancies between the percentage land surface covered by a taxon and its representation in the spectrum may also be explained by reference to relative pollen production by different tree species and their efficiency of dispersal, but this will be considered later in this chapter.

One of the most important contributions which broad geographical surveys of surface pollen can make to our understanding of fossil pollen spectra is the detection of the relative importance of long-distance transport.

There are many records of pollen grains in superficial deposits which belong to species the limits of whose range may be a great distance away. Pollen grains of *Nothofagus*, the southern beech, have been found on Tristan da Cunha, 3000 miles from the nearest tree of that genus. Pollen of *Casuarina*, presumably from Australia, has been found in surface samples from New Zealand, 1000 miles away (Moar, 1969). *Ephedra* has a pollen grain which appears to be capable of wide dispersal and has been found in the modern pollen rain of Ontario, 1300 miles from the known limits of *Ephedra* distribution in the south-western United States (King and Kapp, 1963). Similarly, Tyldesley (1973) has recorded

concentrations of tree pollen (mainly birch and pine) of up to 30 grains per m^3 in air at Lerwick, Shetland, despite the absence of flowering trees on these islands.

Many attempts have been made to trace the movement of pollen and spores in the atmosphere, such as those of Tyldesley, who correlates days with high tree pollen concentrations over Shetland with certain patterns of air movement. In this way he can postulate the trajectory which the pollen grains have followed. For example, an air mass in Shetland on 9th June 1970 contained twenty-nine tree pollen grains per m^3. Fifty-seven hours earlier that air mass had been over Estonia in the eastern Baltic; 39 hr before arrival at Shetland it was over southern Sweden and 21 hr before arrival, over southern Norway. In general, airflow patterns from the east, south-east, south and south-west brought some tree pollen, whereas air movements from the north and north-west brought very little.

Christie and Ritchie (1969) faced a similar problem when they observed anomalous pollen fallout on certain days at Churchill, Manitoba, in the forest tundra region of interior Canada (see Fig. 7.3). The pollen fallout, on the basis of its species composition, was thought to have originated in the aspen parkland area of vegetation. From a consideration of patterns of movement of air masses in the troposphere, Christie and Ritchie came to the conclusion that the movement of this pollen must have taken place in a well-mixed layer of air near the ground. They also observed that there was no rainfall during the first major influx of immigrant pollen, hence Tauber's 'rainwash component' is not the only source of long-distance transported pollen. The mean life of such particles in the troposphere before sedimentation or washout has been estimated at three days.

Hirst and his fellow workers (1967a, b) have made detailed studies of the airborne transport of fungal spores, which are rather smaller than most pollen grains, and also come to the conclusion that most long-distance transport occurs in the turbulent lower layers of the troposphere.

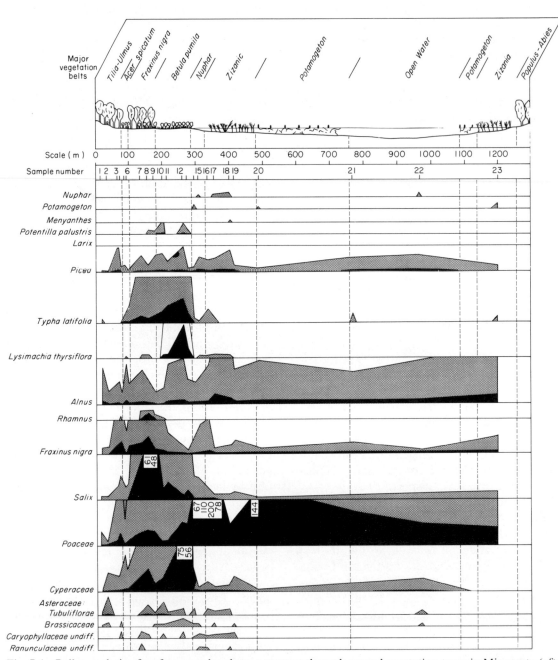

Fig. 7.4   Pollen analysis of surface samples along a transect through several vegetation types in Minnesota (after Janssen, 1966). Only selected pollen types are shown. Dark shaded curves show the proportional representation of each pollen type and light shading shows these values exaggerated × 10.

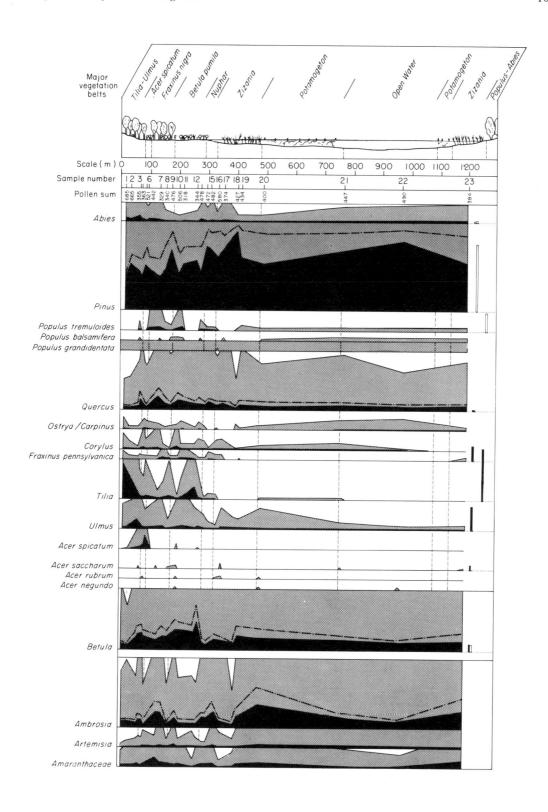

Using aircraft with suction traps they have mapped spore clouds and also studied their vertical structure. They found concentrations of spores of the order 10^4 spores per m^3 at 3000 m. As in the cases already quoted, they too demonstrate that spore clouds can move considerable distances, e.g. from southern France and Spain into Britain, within three days of their origin.

Thus the study of modern pollen fallout patterns leads us to the conclusion that under appropriate meterological conditions light pollen grains and spores may be carried considerable distances from their source of origin and be deposited perhaps thousands of kilometres away. Such a knowledge leads us to considerable caution when interpreting the relative abundance of such pollen and spore types within fossil assemblages.

Studies of surface pollen can also be of assistance in determining the pollen input from local vegetation, especially where surveys have involved intensive study over a small area. Janssen (1966) has conducted surveys of this type in Minnesota. He collected surface samples along a transect 1200 m in length which passed from elm/lime wood through other woodland types, through a zone of emergent aquatic vegetation, across open water and back into woodland on the other side. The relative abundance of selected pollen types along the transect is shown in Fig. 7.4.

Transects of this sort tell us a great deal about the dispersal efficiency of the particular taxa involved, a subject which will be considered in the next section. They also provide a means of estimating how much of the pollen input at a site can be considered to be of immediately local origin. This is seen if the general pattern of pollen proportions over the first 400 m of the transect is compared with that over the remaining 800 m. The first part of the transect is dominated by terrestrial and emergent plants which produce an exceedingly heterogeneous pattern of pollen rain. This is demonstrated by the erratic behaviour of most of the pollen types as one passes along the transect. Evidently the propor-

tions of the various pollen types at each sampling site is largely a function of the proximity of various pollen producing individuals which dominate the spectrum. The pollen input from local sources represents a very high proportion of the total pollen.

In the remainder of the transect, 400–1200 m, there is considerably less fluctuation in pollen proportions from one sampling site to the next, despite the larger horizontal distances between these sites. This part of the transect lies in open water, where the only local pollen producers are floating aquatics. The pollen spectrum is thus dominated by regional pollen, together with some long-distance pollen. Although the floating aquatics do not appear to be influencing the pattern of pollen sedimentation as a result of their own pollen production, they may affect the sedimentation of other pollen types, for example *Larix* pollen appears to be closely correlated with the *Potamogeton* dominated zone.

This study suggests that for lake sites the local pollen component becomes greater as one approaches the lake margin, increases very considerably in regions of emergent vegetation and becomes greater still in semi-terrestrial or terrestrial communities surrounding the lake. Such detailed studies have not been undertaken on ombrotrophic mires, but pollen profiles of domed mires in which *Sphagnum* has often been the dominant vegetation type usually lack any evident dominance of a local pollen component (e.g. Borth Bog, Moore, 1968). On the other hand, blanket mires, which often bear a vegetation of *Eriophorum* spp. (Cyperaceae pollen type) and *Calluna*, frequently show a dominance of these local elements in their pollen profiles (e.g. the Welsh blanket peats, Moore 1968, 1973).

In this section we have considered the possible geographical origins of the pollen which arrives at the locus of its sedimentation and fossilization. In only a very few cases is it possible to guess at the geographical source of a particular pollen grain, but from a consideration of the evidence of stratigraphy, macrofossil content of the peat and the nature of the catchment site, it is sometimes possible to estimate the relative importance of local, regional and long-distance pollen components in the fossil assemblage. Such an estimate is greatly assisted by a knowledge of present-day fallout patterns of pollen both over wide geographical areas and within various habitats and plant communities.

Dispersal efficiency

In any consideration of the geographical origins of the components of fossil pollen assemblages, it is evident that different species behave in different ways. Thus a comparison of Tables 7.1 and 7.2 shows that pollen grains are found outside the geographical range of the genus, but others, such as *Quercus* do not appear in any quantity in the pollen spectrum of communities not containing the genus. The reason for this differential behaviour between taxa lies largely in the dispersal efficiency of their pollen grains and also in the stature of the plants themselves.

The greater part of the non-local element in any pollen assemblage consists of the pollen of anemophilous (wind-pollinated) species. The pollen grains of anemophilous plants generally have a higher dispersal efficiency in air than those of species using other pollination mechanisms. It is not always possible, however, to account for differential representation at a site by reference to differential transport in air. Despite the fact that conifer pollen is well represented outside the geographical limits of conifer trees, Firbas (1934) was unable to demonstrate greater flotation potential of conifer pollen in still air when compared with angiosperm pollen, such as alder. Evidently, if the bladders on these grains do assist flotation in air, the advantage is offset by the greater size of conifer grains. Firbas concluded that over-representation by conifers must be due to greater pollen production rather than dispersal efficiency. The two are closely interrelated.

Erdtman (1969) has calculated the volume, weight and settling velocity in still air of a number of pollen types. They are shown in Table 7.3. However, it is doubtful whether these data are of any great value in assisting interpretation of pollen spectra, because of the complicating effect of additional factors such as pollen production and the position above the ground where the pollen is released and enters the moving air mass.

TABLE 7.3 *Volume, weight and settling velocity in still air for a number of pollen types (from Erdtman, 1969).*

Pollen type	Volume, m^3	Weight, $g \times 10^{-9}$	Settling velocity cm s^{-1}
Picea	132 000	72·8	6
Fagus	51 770	37·0	5
Pinus	47 030	18·4	3
Corylus	10 150	10·2	2·3
Alnus	9 070	6·8	1·6
Betula	7 540	6·1	1·5
Taxus	7 130	4·1	1·0
Juniperus	9 460	3·8	0·9

The low dispersal efficiency of species which are not wind pollinated may occasionally be of value in determining the local element within a pollen assemblage. Entomophilous pollen grains are liable to be of local origin. Once again it is impossible to isolate this characteristic from the low pollen production associated with insect pollinated plants. Thus when Proctor and Lambert (1961) examined the surface pollen rain associated with *Helianthemum* communities, they found that even high densities of *Helianthemum* could result in a low representation in the pollen rain. This is a result of the combined effects of low pollen production and poor dispersal efficiency.

Obviously these differential dispersion effects will complicate the business of interpreting pollen spectra. As Oldfield (1970) points out, the only circumstances in which one can make direct inference of vegetational change is the situation where one uniform anemophilous dominated vegetation type is being replaced by a similar vegetation type. A situation of this sort is described by Hafsten (1961) in the south-western part of the United States where pine forest replaces *Artemisia* steppe.

A further problem when considering the differential dispersal efficiency of pollen types is that of impaction upon solid obstacles. This factor is of particular importance within the trunk space of forests, as has been demonstrated by Tauber (1967b). Here, pollen grains are literally filtered from the moving air mass by impaction upon trees and shrubs. Tauber also claims that the capture rate will vary for different pollen types; large grains such as *Fagus*, *Ulmus* and *Tilia* will be three or four times more likely to be caught by obstacles than small grains such as *Betula*, *Corylus* and *Alnus*. The filtering effect of obstacles in this way has considerable biological

provide yet one more restriction upon pollen movement which acts in a differential manner.

Chamberlain and Chadwick (1972) have studied pollen impaction upon various surfaces using radioactively labelled pollen grains. They found that if surfaces (such as twigs and leaves) were moist, the degree of retention of impacted pollen grains was much higher than for dry surfaces. Also the efficiency of pollen capture becomes greater with increasing windspeed.

Pollen productivity

Intricately linked with the problems posed by the dispersal efficiency of pollen grains are those associated with differential production of pollen.

TABLE 7.4   *Pollen production of various species expressed in different ways. The index of relative pollen production in the final column is based upon estimates of the pollen production of an individual over a period of 50 years and is expressed relative to beech (estimated production $2{\cdot}45 \times 10^{10}$ pollen grains). This value is taken as unity (after Erdtman, 1969).*

	No. of pollen grains per anther	No. of pollen grains per flower	No. of pollen grains per catkin	Index of relative pollen production ($Fagus = 1{\cdot}0$)
Trifolium pratense	220	—	—	—
Acer platanoides	1 000	8 000	—	—
Malus sylvestris	1400–6250	—	—	—
Calluna vulgaris	2000 tetrads	—	—	—
Fraxinus excelsior	12 500	—	—	—
Secale cereale	19 000	57 000	—	—
Rumex acetosa	30 000	180 000	—	—
Juniperus communis	—	400 000	—	—
Pinus sylvestris	—	160 000	—	15·8
Picea abies	—	600 000	—	13·4
Betula pubescens	—	—	6 000 000	—
Alnus glutinosa	—	—	4 500 000	17·7
Quercus robar	—	1 250 000	—	—
Fagus sylvatica	—	—	175 000	1·0
Quercus petraea	—	—	—	1·6
Carpinus betulus	—	—	—	7·7
Betula pendula	—	—	—	13·6
Corylus avellana	—	—	—	13·7
Tilia cordata	—	—	—	13·7

significance, since the ultimate function of an airborne pollen grain is that it shall collide with and adhere to a stigma of the same species. Its effect as far as palaeopalynology is concerned is to

Table 7.4 gives various data concerning pollen production in a number of species (from Erdtman, 1969). It can be seen that entomophilous species (e.g. *Trifolium*, *Acer*, *Malus* and

Calluna) have considerably lower pollen production per anther than anemophilous species. However, even within the anemophilous group there is considerable variation in pollen production, especially when expressed on a long-term basis in relation to beech. *Alnus* has the highest productivity on this basis, followed by *Pinus*. *Fagus* has the lowest productivity of those measured.

Andersen (1967) has made similar studies on the pollen productivity of trees, but has used surface pollen samples from beneath mixed deciduous forest in order to obtain indices of productivity. He collected moss polsters from the woodland and recorded the trunk area of all tree species within a 30 m radius of the sample. He was then able to produce graphs in which he related the basal area of each tree species to its proportional representation in the pollen sum. He conducted linear regression analyses and obtained a correlation coefficient, r, for each species. From the regression equations he was able to produce relative production indices, relating them to beech, which he took as unity. His values were as follows (Erdtman's figures in parentheses):

Betula	4·1	(13·6)
Quercus	4.1	(1·6)
Tilia	0·7	(13·7)
Alnus	1·5	(17·7)
Fraxinus	0·6	(—)

These figures can be seen to differ very considerably from Erdtman's data. *Alnus*, for example, is considerably lower than the value obtained by Erdtman, whereas *Quercus* is higher. Andersen's data are probably more useful when interpreting fossil pollen assemblages from forested areas, since they effectively combine the factor of pollen production with short-range dispersal efficiency, i.e. over a range of 30 m.

Andersen suggests that these figures can be used for the adjustment of relative proportions in the pollen diagrams. However, so many other factors influence the curves on pollen diagrams that even if these indices of productivity were perfect estimates (which they are probably not),

adjustment of the curves on the basis of this one source of variation could cause more confusion than clarification. It is valuable, however, to bear such figures as these in mind when interpreting pollen assemblages, particularly if there is strong local influence from trees.

Local environment

If one accepts Tauber's model of pollen dispersion (Fig. 7.2), then the distance over which a pollen grain is transported and hence the proportional representation of a taxon within a fossil assemblage, varies with the nature of the air stream into which the grain is initially released. Thus a plant dwelling beneath a woodland canopy will release its pollen into the trunk space air mass and the likelihood of its escaping into the canopy air stream or to higher levels will depend upon the turbulence of the air masses. For this reason, undershrubs and herbs beneath a closed woodland canopy will not be distributed as widely as those types released above the canopy. These taxa will tend to be under-represented in fossil assemblages in terms of the ratio between grain frequency and the abundance of that taxon in the area.

The situation with regard to canopy trees will be yet more complex. Pollen emitted from flowers on the upper part of the canopy will enter the air stream above the canopy and will experience a wide dispersal. Pollen from flowers borne on lower branches is liable to sediment downwards in these relatively still conditions (see Fig. 7.5) and hence will enter the trunk space air mass. If, however, one considers a single mature tree which is not surrounded by other trees, but stands on its own in an exposed position, then all of the pollen released by it will enter the turbulent air streams which will carry the pollen over greater distances. As a result, the pollen from such a tree is more likely to be represented in a fossil assemblage.

It is particularly important to bear this point in mind when considering a series of assemblages which corresponds to a period of forest clearance. If clearance is of a type which leaves occasional trees in a park-like environment, then the pollen rain of that particular species could be increased because of the better facilities for dispersion. Troels-Smith has observed this effect within lake basins in Switzerland.

Many of the high values for the pollen of tree species obtained from surface samples in areas of intensive human activity could well be explained by this dispersal characteristic of isolated trees. This rather artificial state of affairs makes it particularly difficult to relate proportional, or even absolute, pollen values in surface samples to the abundance of the respective species in the region round about. This, together with the difficulties involved in knowing the distance of the collection site from the pollen source (see Oldfield, 1970), severely limits the value of simple ratios relating pollen abundance to land area occupied as a tool for the interpretation of pollen assemblages (see Davis, 1963).

Time of flowering

Godwin (1934) agreed with Erdtman (1931) that hazel pollen, since it is produced very early in the year, could arrive on bog surfaces while they are still frozen and could decay before being incorporated into the peat. As a result, they believed,

it is possible for considerable quantities of hazel to be present in an area without leaving any record in the peat profile. This viewpoint can scarcely be regarded tenable nowadays, but the early flowering of hazel and other species could have other important consequences.

Fig. 7.5 shows the wind speed profiles obtained by Geiger (1965) for a deciduous forest before and after leaf emergence. From this it can be seen that following leaf emergence the canopy is a region in which there is little air movement, which means that there is little likelihood of transfer from the trunk space to the air above the canopy at this stage. Before emergence, however, there is a gradual decrease in wind speed as one moves downwards through the forest profile, which increases the incidence of mixing between the air masses. Also, for any given above-canopy wind speed, the wind speeds within the forest profile are higher prior to leaf emergence than they are afterwards; this also leads to greater air mixing. From these microclimatic data it can be seen that the flowering of understorey shrubs before the canopy opens will allow greater dispersal of pollen than would be the case if flowering were delayed until after the opening of the leaf canopy.

In terms of pollen diagram interpretation, these considerations lead to the conclusion that the time of flowering is effectively a further variable which can lead to differential representation within a pollen spectrum. Early flowering species, such as hazel (*Corylus*) and ivy (*Hedera*) are more likely to be dispersed beyond the canopy than late-flowering species such as honeysuckle (*Lonicera*).

Morphological and physiological condition

It is, perhaps, stating the obvious to say that pollen analysis can take into account only those taxa which are capable of flowering. It is possible, however, for a species to be present, perhaps even abundant, in an area without possessing that capacity.

Fig. 7.6 shows a hypothetical performance

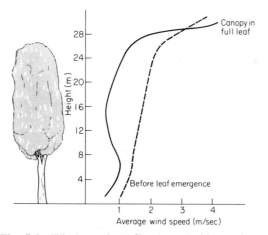

Fig. 7.5 Wind speed profiles in a deciduous forest before and after leaf emergence (after Geiger, 1965).

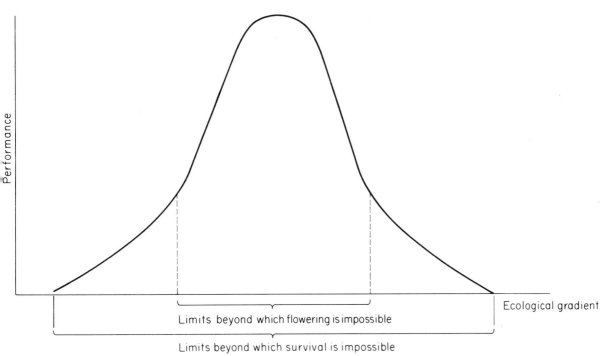

Fig. 7.6 Idealized curve relating the performance of a species to an ecological gradient and showing limits beyond which flowering and physical survival respectively become impossible.

curve for a species in relation to an environmental gradient. The curve is idealized in several respects, particularly in that it does not take account of the possibility of competitive exclusion in situations where the species is under physiological stress. It does, however, demonstrate that a species can occur beyond the potential limits of flower production. The maintenance of a population in such a situation depends upon a number of factors:

1. The species must be capable of vegetative proliferation in this situation.
2. Population replenishment by seed invasion must be possible.
3. Germination and establishment of these replacement seeds must be possible from time to time, though replenishment of this sort need not necessarily be a frequent occurrence; it depends upon the rate of mortality.

Given these conditions, a moderately healthy population of a species can be maintained be-

yond the normal tolerance limits for reproduction in the species. Pearsall (1950) provides an interesting example of this situation in *Juncus squarosus*, the heath rush, which can survive in a vegetative state at altitudes where it is incapable of producing viable seeds.

A situation in palynology where this explanation could be invoked is afforded by the behaviour of *Juniperus* at the close of the Devensian glaciation. During the latter part of both Younger Dryas and Older Dryas periods in many late-Devensian sites in Britain there is a very pronounced and sudden rise in *Juniperus* pollen (Pennington *et al.*, 1972; Moore, 1970). Pennington and Bonny (1970) have shown this rise at the close of the Older Dryas in Blelham Bog, Lancashire, to be accompanied by a large absolute rise in pollen input from many species. Either this sudden rise was the result of rapid immigration on the part of juniper, or it resulted from a climatic threshold being crossed which permitted flowering whereas this was not possible previously.

There are other factors which might prevent plants from flowering apart from the limitations imposed by physical conditions. One such factor is grazing and another is the effect of man lopping off flowering shoots. The latter explanation has been put forward as a possible way by which man might have caused the elm decline in northwest Europe at around 3000 BC.

Morphological and physiological condition is therefore an important consideration in the interpretation of pollen diagrams.

Density

At the very heart of the theory of pollen analysis is the concept that the frequency of a pollen type within an assemblage is a function of the abundance of that taxon in the surrounding area. It was this hope which inspired the original researches of von Post (1916) and it is this belief which finally permits us to begin the reconstruction of palaeovegetation on the basis of a fossil pollen assemblage. This reconstruction, however, can begin only all other sources of variability (discussed in the previous seven sections) have been taken into consideration, and then only when we have some idea of the overall relationship between pollen input and the abundance of the species in the region.

Davis (1963) has proposed a ratio which she designates R to represent this relationship, where

$$R_A = \frac{\text{Percentage pollen of species A}}{\text{Percentage representation of species A in the vegetation}}$$

The R value effectively takes into account differences between species in terms of their dispersal efficiency and pollen production so that, in Davis' opinion, the true proportional representation of a species in vegetation can be determined if the pollen percentage is 'corrected' by using R values. Although this approach would begin to place pollen diagram interpretation on a level slightly above the intuitive, it does not account for all sources of variation. In particular, the local environment of the pollen producing

individual affects the R value for the species, as will the distance between the pollen source and the site of accumulation. Since these two variables cannot be inferred from the pollen assemblage itself, the R value may be positively misleading to the interpreter; at best it may lull him into a false sense of security.

The alternative to the Davis R value is a return to intuitive interpretation. Although no one can regard this as satisfactory, it does allow the interpreter to bear in mind all possible sources of error and to qualify his interpretation accordingly (see Chapter 8). In this respect, as Oldfield (1970) has pointed out, the only hope presenting itself to the pollen analyst which may allow him to place his interpretation on a more objective basis is that the pollen assemblage can be split into components derived from different plant communities. If something is known of the nature and composition of the plant communities contribution to the pollen rain, then the problem of distance from pollen sources could possibly be resolved and that of the structural nature of the pollen-producing vegetation may also be open to resolution.

Plant communities

Some time ago West (1964) pointed out that the pollen assemblage does not normally represent an even mixture of pollen types derived from an homogeneous vegetation. Rather does it comprise a collection of pollen from a complex mosaic of plant communities or, depending upon one's philosophy of vegetation, a multidimensional continuum of plant individuals of various species. If one could determine something of the constitution of these communities, then this, when considered in the light of local topography and geology around the site of accumulation, could permit some degree of spatial vegetational reconstruction.

As in the study of phytosociology, one of the most useful criteria upon which communities can be recognized is the species of narrow ecological tolerance which is strictly limited both in the environmental conditions under which it grows and also in the accompanying species with which it is normally associated. Occasionally such species have a distinctive pollen or spore morphology which allows a specific determination to be made; one can then make fairly firm assertions concerning the vegetation in which the species is found.

One example of this situation is afforded by *Polemonium caeruleum* which has a distinctive pollen grain (Plate 14) often found in British sediments of late Devensian age. This species is strongly calcicolous and is also confined to a characteristic tall herb plant community described by Holdgate (1955). Some of the other taxa found in such a community are also frequently represented in the *Polemonium* pollen assemblage and this also aids the process of recognizing a plant community within a pollen assemblage.

A further example is that of *Artemisia norvegica* which is again recognizable on the basis of its pollen morphology and has been found at a few late-Devensian sites in Britain, e.g. Moore (1970) in central Wales, and Birks (1973a) on the Isle of Skye, Scotland. The species is very restricted in its present-day distribution in Britain, but where found it is in a characteristic plant community dominated by *Juncus trifidus* (McVean and Ratcliffe, 1962). It is tempting, though perhaps a little risky, to assume that it was associated with similar plant communities in late Devensian times when it enjoyed a far wider distribution.

It is possible, therefore, on the basis of the recognition of such indicator species as these, to attempt the reconstruction of the contemporary mosaic of vegetation types from a fossil pollen assemblage. It must be clear, however, from the foregoing discussion of the multitude of factors which influence the representation of a species in such an assemblage, that such reconstruction must inevitably be tenuous. Evidence from other sources, particularly macrofossils, is always of considerable value in the confirmation or modification of vegetational reconstruction based on pollen and spores.

The second assumption

The first assumption involved in the interpretation of a pollen diagram was that the pollen assemblage reflects the surrounding vegetation. The preceding section has shown that this assumption is true in principle, but that the 'reflection' is far from straightforward and many modifying factors need to be considered constantly. The second assumption is that vegetation reflects environmental conditions, and this assumption will now be analysed briefly.

There are two distinct considerations which underlie the application of this assumption to palaeoecological data. The first consideration concerns the way in which factors of the environment may influence the performance or distribution of a plant species in such a specific manner that the very presence of the species documents the precise environmental situation. The second consideration concerns the application of such a knowledge of the current behaviour of a taxon to a fossil taxon which is morphologically indistinguishable. Both of these considerations merit careful attention because neither can be fully justified.

A great deal of modern plant ecological research is concerned with the responses of species to the variable factors of the environment. The palaeoecologist essentially leans upon this work in order to be in a position to interpret his reconstructed plant assemblages in terms of their environment. As a result of these researches it is now believed that each species of plant has an optimum requirement for any given ecological variable and also it has limits relating to that variable beyond which it is unable to survive (see Fig. 7.6). These theoretical limits, however, may not be achieved in practice because of competition from more efficient organisms or because of sensitivity to disease or predation when approaching its tolerance limits. Any given species will have a variety of optima and tolerances for the whole spectrum of environmental variables which affect its establishment and growth. If any of these limits are approached in any particular habitat or geographical locality then that factor may be the determinant controlling the presence or absence of the species at any given site. In other words, the factor in least supply will limit the distribution of the species in that area (assuming there are no physical barriers, e.g. seas, mountain ranges, extensive unsuitable habitats, etc. which are influencing dispersal potential). This is basically a restatement of Liebig's law of limiting factors, which is fundamental to any understanding of an organism's response to its environment.

When considering plant distributions, however, it is necessary to bear in mind that different factors may be limiting distributions when one considers the species at different spatial scales (see Cox, Healey and Moore, 1976). For example, on a global scale a species may be limited by its climatic requirements, whereas when a smaller area is considered on a larger scale it may be limited by a soil factor, e.g. pH, nutrient availability or water regime. If a map is constructed for a species on the basis of its fossil occurrences at a particular time, then its interpretation must take into consideration the scale at which the distribution pattern of the species is being represented.

If data are available, then, concerning the environmental requirements of a species at the present day, these data can be used to adduce palaeoenvironments. The greatest problem, however, particularly with climatic data, is to know which component of the overall environment in fact limits distributions. Some examples may illustrate this point. Dahl (1951) has attempted to define the precise component of climate which limits the distribution of montane plant species in Norway. He found that the limits of continuous distribution of many montane species coincided with certain isotherms of mean annual summer temperature. Although one must be cautious when taking a correlation to imply causation, it does provide a possible clue to the climatic factor which may be limiting certain montane plants. The next step in the enquiry involved experimentation, and Dahl found, as he had predicted, that high summer temperatures

were harmful to many montane species. This explained why many such species are able to grow at low altitudes in western Norway, where oceanic conditions keep summer temperatures low.

Obviously, different species appear to be limited by different mean annual summer maxima; some examples are:

Sedum rosea	25°C
Salix herbacea	26°C
Dryas octopetala	27°C
Koenigia islandica	24°C

Conolly (1961) has shown, however, that latitude is important in determining the limits for a species and suggests, for example, that the distribution of *Salix herbacea* is more closely correlated with a mean annual summer maximum of 23°C in Scotland and Ireland and 25°C in England. Armed with such a knowledge of the requirements of certain species, many of which can be identified on the basis of pollen alone, it then becomes possible to reconstruct palaeoclimates on the basis of fossil distributions.

When Dahl and Conolly originally put forward their ideas, there were too few fossil records for them to attempt climatic reconstructions on any wide scale. Since then, however, many palaeoecological investigations have taken place in the British Isles and these data have been collected together by Conolly and Dahl (1970) in an attempt to define the temperature conditions within the British Isles during Devensian, late-Devensian and Flandrian times. Their data collation allows them to conclude that conditions during the full Devensian glacial were at least 5°C colder than today in East Anglia and 3·5–4°C colder in the English Midlands. They regard the late-Devensian as being on average 3°C cooler than today.

One major problem associated with the use of plants as palaeoclimatic indicators is that they respond rather slowly to climatic changes. If climate changes rapidly, as it did during late-Devensian times, then the vegetation may never attain a state of equilibrium with the climate.

Under these conditions, climatic extrapolations from botanical evidence may be quite erroneous, as has been demonstrated by Coope (1970) for the British late-Devensian. He has analysed sediments for beetle remains and has discovered that sediments which have often been ascribed to the latter part of pollen zone I (the Older Dryas), which has been considered on botanical grounds to have been formed during sub-arctic conditions, often contains the remains of beetles which are now typical of more southerly latitudes. In other words, if this period were one in which the climate was warming up very rapidly, the beetles, being more mobile, could migrate and take advantage of new opportunities long before the plants. On the evidence of these mobile organisms one may conclude that this period was in fact warmer than the present day. Thus botanical evidence has severe disadvantages when reconstructing palaeoclimates in times of rapid change and instability.

During Flandrian times there are further problems in the use of plants as climatic indicators. For example, Iversen (1944) regarded mistletoe (*Viscum*), ivy (*Hedera*) and holly (*Ilex*) as useful climatic indicators because of their frost sensitivity. Troels-Smith (1960) was later able to show that two of these, *Viscum* and *Hedera*, were important fodder plants for prehistoric cultures and therefore their decline in the later Flandrian does not necessarily mean a deterioration in climate.

It is not always climatic factors, therefore, which limit the distributions of plants. Soil factors have already been mentioned as important determinants within a plant's gross climatic limits. It is therefore possible to use palynological data as a means of determining past soil conditions in an area. To quote an example which has already been used in a different context, the presence of *Polemonium caeruleum* is a sure indication of the presence of calcium carbonate in a soil. Occasionally this species has been recorded from late-Devensian sediments at sites which are now extremely calcium deficient, e.g. those on the Ordovician and Silurian shales of

mid-Wales at Rhosgoch, Radnorshire (Bartley, 1960) and Elan Valley, Cardiganshire. Presumably the freshly derived and unleached, immature soils derived from the recent glaciation were sufficiently basic for the survival of such a demanding species.

The presence of some pollen types, particularly during the later phases of the Flandrian, is interpreted as indicating the activities of man in the vicinity. Perhaps the most useful type in this respect is *Plantago lanceolata,* the pollen of which can be determined to the level of species (see page 54). Often other pollen types associated with weed species are found at the same levels in the profile (Godwin, 1960). Turner (1964a) proposed that a pastoral/arable ratio could be estimated of the basis of the composition of the weed flora; *Plantago lanceolata* and *Rumex acetosa* are associated mainly with pastoral conditions, whereas *Plantago major* and *Artemisia* are found more frequently in arable situations. The presence of cereal pollen is also helpful in the determination of the importance of arable agriculture.

It does appear, therefore, that conclusions can be reached concerning the environmental conditions which obtained at the time when a pollen assemblage accumulated. The precision with which such conclusions can be made is dependent largely upon the information available regarding the present-day requirements of the species in question.

There is a final consideration which must be made and which must be kept in mind throughout the process of interpreting fossil assemblages. It has been assumed so far that the behaviour, physiology and environmental requirements of the fossil species were identical to those of its present-day morphological counterparts. This assumption is generally held to be valid and is often regarded as one of the great advantages which the Quaternary palaeoecologist has and which the student of extinct taxa lacks. However, one must always admit the possibility that in certain instances the assumption may be unfounded.

The identification of fossil material is based

entirely upon morphological and anatomical comparisons. When these characters compare closely with those of modern taxa, we are contented to confirm identifications. However, morphological identity does not necessarily imply physiological identity. There are many known cases in which a modern taxon has ecotypes which are closely similar in morphology and yet differ widely in physiology. A very good example is that of *Festuca ovina*. This species is found growing both on chalky and on acidic substrates, the two populations being morphologically similar. Their ability to tolerate such widely differing substrates has been shown by Snaydon and Bradshaw (1961) to be due to considerable differences in the ion-absorption properties of the two ecotypes.

When considering a species over a long period of time, during which environmental conditions are changing most profoundly (as they have during the past 10 000 years), one must regard it as possible that ecotypic variation of this type may have occurred in many species. It is also possible that selective pressures which have varied at different times in the past will have selectively favoured different ecotypes. If this is so then a single morphological taxon may have played a number of ecological and physiological roles during a period of rapid environmental change.

On very few occasions has this argument been called upon in order to explain an anomalous situation. One such situation is the behaviour of *Pinus sylvestris* in the mid post-glacial, when it appears to increase in north-west Britain during a period of apparent increased climatic wetness. Oldfield (1965) considers it possible that at least two ecotypes of pine existed in Britain at that time which responded differently to the climatic conditions. Green (1968) resorts to a similar

argument to explain the former abundance of *Sphagnum imbricatum* in Britain as a major peat-forming species when it is now scarce and grows in isolated hummocks.

Conolly (1961) has stressed the importance of using all available taxa in the reconstruction of past environments rather than relying solely upon one or two 'indicator types'. This would appear to be the only safeguard against misinterpretation of fossil assemblages as a result of physiological changes which have occurred in key types during the course of time. Such changes are unlikely to have occurred in the same physiological direction in a number of taxa at the same time.

Conclusion

In this chapter, we have examined all of the factors which influence the production, transport and sedimentation of pollen and spores, and which therefore modify the constitution of a fossil microspore assemblage. Despite all the commendable attempts to place the interpretation of pollen diagrams upon an objective plane, it remains largely an intuitive process, the precision of which depends strongly upon the experience and degree of bias found in the palynologist. Further research into the documentation of current pollen spectra in different habitats will make the task of interpretation easier, as will the accruing knowledge of the ecological and environmental requirements of the species involved. It remains to be seen, however, whether the collation of such information will ever permit objective, mathematical analysis of pollen diagrams in which all of the many variables can be considered at the same time. We are at present very far from this aim.

Chapter 8
The Future of Pollen Analysis

This is an age in which the cross-fertilization of ideas between scientific disciplines is resulting in an accelerating rate of development in many fields. Palynology is constantly drawing upon and benefiting from advances in such fields as microscopy, cytology, epidemiology, meteorology, ecology, statistics and mathematics. Nevertheless the basic techniques and principles of pollen analysis are simple and easily learned; the observations of unskilled practitioners rapidly attain a degree of reliability, as a result of a little experience. Simple, unsophisticated analyses of peat and lake sediments can still reveal new evidence concerning the past history of our vegetation and the environment in which it developed. One of the prime aims of this book is to encourage such simple yet essential work by providing a means of extracting, identifying and interpreting subfossil Quaternary pollen and spores.

Despite the basic simplicity of the techniques of pollen analysis, many questions can be answered only by using more sophisticated approaches. For example, more accurate identification of pollen types can lead to a higher degree of resolution and precision in the examination of past environments. In this field one must look to higher magnification microscopy. The transmission electron microscope offered some hope of solution for certain problems in identification, for example in the separation of the British species of *Tilia* (Chambers and Godwin, 1961). However, on the whole the transmission electron microscope proved a disappointment to palynologists, apart from those concerned with developmental aspects of structure in pollen exines (see Heslop-Harrison, 1968, and Heslop-Harrison and Dickinson, 1968). The construction of pollen replicas did provide some opportunity

for surface studies (see, for example, Yamazaki and Takeoka, 1962).

The scanning electron microscope has proved a much more useful tool for palynologists (see Leffingwell and Hodgkin, 1971). This has allowed pollen analysts to take a much more detailed look at the sculpturing of pollen exines, as is shown by some of the photographs in this book. Chambers and Godwin (1971), returning to the problem of the differentiation of *Tilia* species, employed scanning electron microscopy to much greater advantage. Ferguson and Webb (1970) have used the fine sculpturing details revealed by the scanning electron microscope in their studies on the genus *Saxifraga*. They found that their classification of 105 species into groups on the basis of pollen grain fine sculpturing corresponded well with the existing classifications based on the morphology of the plants. With the scanning electron microscope to illustrate their sculpture, pollen grains may become an even more important tool to the taxonomist than Erdtman (1963, 1969) has suggested.

In the identification of pollen grains from deposits the scanning electron microscope might have a more limited use because so many identifications depend on structural details of the pollen exine which are not visible in surface view. It is conceivable that it could be useful where the separation of pollen types does involve some important surface feature, such as the difficult distinction between *Corylus* and *Myrica* (see page 53). Hebda and Lott (1973) have used the scanning electron microscope to show the variability of sculpturing which can exist within the pollen grains of one species. They found that the sexine sculpture in *Saintpaulia* pollen grains varied according to the temperature at which the plant was grown. They emphasize that, although this may be a disadvantage when trying to distinguish between the pollen grains of two similar species, sculpturing changes found in an identifiable pollen type which is present all the way up a sediment core may well reflect minor climatic changes.

Besides microscopy, there are other techniques which have permitted a more accurate identification of pollen types. Numerical taxonomy is an area where objective measurements have replaced subjective judgments and this approach has proved particularly valuable in elucidating vexed palynological problems. For example, Birks (1968) provided numerical criteria based on pollen grain size and pore depth which allowed the separation of *Betula nana* from other species of *Betula*. He has since used similar techniques for the separation of types within the genus *Isoetes* and *Lotus* in late-Devensian deposits (Birks, 1973a).

Pollen morphology and taxonomy are among the most fundamental aspects of palynology, but Quaternary pollen studies have also been assisted very considerably in recent years by the use of numerical methods of analysis to aid in the zonation and interpretation of fossil pollen assemblages. Initially, this began with the application of computer techniques as labour-saving devices for the analysis of raw data and the construction of pollen diagrams (see, for example, Squires and Holder, 1970). Since then, techniques of multivariate analysis have been developed which allow pollen diagrams to be zoned on mathematical criteria (see Chapter 6, also Birks, 1974) and which provide methods for comparing one pollen diagram with another (Gordon and Birks, 1974). Perhaps the most important outcome of such research has been the way in which it has confirmed most of the earlier, intuitive judgments of experienced palynologists lacking those tools. This is not really surprising since the human mind is capable of assessing and comparing data with considerable facility. Although the mind does not normally operate on the basis of prescribed mathematical formulae, it does weigh a number of considerations with great flexibility.

This flexibility is normally lacking in computer-based analyses.

Birks (1974) has shown that numerical zonation of British pollen diagrams usually produces results which are consistent with traditional, intuitive zonations. The one exception is the zone boundary based upon the elm decline, which is frequently neglected by numerical techniques. Here is a pollen change of considerable supposed palynological interest which, because it results in only small changes in certain 'critical' pollen types (e.g. *Ulmus, Plantago lanceolata, Pteridium*), does not make an impression upon the inflexible computer, programmed to respond only to certain types of stimuli. Perhaps it is at this point, where value judgments concerning the importance of certain pollen characteristics become involved, that palynology ceases to be a science. It is in the same way that the historian, in sifting through masses of data, picks upon certain events and certain words as being of critical importance in influencing the course of history. Such estimations of importance cannot be proved mathematically, but can usually be defended by logical argument. This is frequently the position of the palynologist.

Numerical techniques do have the great advantage that they can assist the inexperienced in coming to the right conclusions, though even here he may fail to choose the most appropriate numerical technique. The important point is not to overestimate the power of numerical methods; they are a valuable tool, but they are not foolproof and neither are they indispensable. In trying to recommend palynology as a technique for those in schools and colleges without access to sophisticated machinery or palynological experience, we seek to emphasize that pollen analysis, identification and interpretation can be achieved satisfactorily without any such recourse.

All the variables which must be taken into consideration during the interpretation of pollen diagrams were discussed in Chapter 7. The picture given there may appear a little bleak because the variables are so numerous and our knowledge of their operations is so poor. However, despite these problems the broad conclusions from pollen analytical studies appear to have considerable validity. The recent spate of surface pollen studies (e.g. Lichti-Federovich and Ritchie, 1968) has provided a good body of information upon which to base the interpretation of fossil assemblages of pollen. Much more data is still needed to permit more detailed reconstruction of past vegetation types, but there does appear to be a broad correspondence of regional vegetation with a characteristic surface pollen spectrum. Once again, the intuitive probings of the fathers of palynology have proved, rather fortuitously, to be adequate.

In conclusion, modern studies and more sophisticated techniques have presented us with a much clearer picture of the pitfalls and problems associated with the identification and interpretation of fossil pollen than was available to the early workers in this field. Despite this, their basic conclusions have generally been confirmed, which suggests that pollen analysis is a robust and flexible tool. Until very recently the technique has been available to only a few specialists and has been taught in a relatively small number of university departments. It is both the attractiveness of the technique and its far-reaching implications which have led to a demand for a wider dissemination of its practical aspects. This demand has come from specialists in non-botanical disciplines such as archaeology and pedology, as well as from school biology courses, where the results of palynological research may be known but where practical experience has been denied because of the inacessibility of a laboratory manual. It is in the hope of filling these gaps and encouraging the development of palynological studies in these situations that this account has been brought together.

Bibliography

Afzelius, B. M. (1956) Electron microscope investigations into exine stratification. *Grana Palynol.* 1, 22–37.

Andersen, S. Th. (1965) Mounting media and mounting techniques. In *Handbook of Palaeontological Techniques*, ed. Kummel, B. G. and Raup, D. M., Freeman, San Francisco, pp. 587–98.

Andersen, S. Th. (1967) Tree pollen rain in a mixed deciduous forest in South Jutland (Denmark). *Rev. Palaeobotan. Palynol.* 3, 267–75.

Andersen, S. Th. (1960) Silicone oil as a mounting medium for pollen grains. *Danm. geol. Unders. Ser. IV*, 4 (1), 1–24.

Andrew, R. (1971) Exine pattern in the pollen of British species of *Tilia*. *New Phytol.* 70, 683–6.

Banerjee, U. C. and Barghoorn, E. S. (1971) The tapetal membranes in grasses and Ubisch body control of mature exine pattern. In *Sporopollenin*, ed. Brooks, J. B., Grant, P. R., Muir, M., Van Gijzel, P. and Shaw, G. Academic Press, London and New York, pp. 126–7.

Barkeley, F. (1934) The statistical theory of pollen analysis. *Ecology* 15, 283–9.

Bartley, D. D. (1960) Rhosgoch Common, Rads., stratigraphy and pollen analysis. *New Phytol.* 59, 238–62.

Bartley, D. D. (1967) Pollen analysis of surface samples of vegetation from arctic Quebec. *Pollen Spores* 9, 101–6.

Beer, R. (1911) Studies in spore development. *Ann. Bot.* 25, 199–214.

Benninghoff, W. S. (1962) Calculation of pollen and spore density in sediments by addition of exotic pollen in known quantities. *Pollen Spores* 4, 332–3.

Berglund, B. E. (1973) Pollen dispersal and deposition in an area of Southeastern Sweden—some preliminary results. In *Quaternary Plant Ecology*, ed. Birks, H. J. B. and West, R. G. Blackwell, Oxford, pp. 117–30.

Berglund, B., Erdtman, G. and Praglowski, J. (1960) On the index of refraction of embedding media and its importance in palynological investigations. *Svensk, bot. Tidskr.* 53, 452–68.

Birks, H. H. (1970). Studies in the vegetational history of Scotland. I. A pollen diagram from Abernethy Forest, Inverness-shire. *J. Ecol.* 58, 827–46.

Birks, H. H. (1972a) Studies in the vegetational history of Scotland. II. Two pollen diagrams from the Galloway Hills, Kirkcudbrightshire. *J. Ecol.* 60, 183–217.

Birks, H. H. (1972b) Studies in the vegetational history of Scotland. III. A radiocarbon dated pollen diagram from Loch Maree, Ross and Cromarty. *New Phytol.* 71, 731–54.

Birks, H. J. B. (1968) The identification of *Betula nana* pollen. *New Phytol.* 67, 309–14.

Birks, H. J. B. (1970) Inwashed pollen spectra at Loch Fada, Isle of Skye. *New Phytol.* 69, 807–21.

Birks, H. J. B. (1973a) *Past and Present Vegetation of the Isle of Skye*. Cambridge University Press, London.

Birks, H. J. B. (1973b) Modern pollen rain studies in some arctic and alpine environments. In *Quaternary Plant Ecology,* ed. Birks, H. J. B. and West, R. G. Blackwell, Oxford, pp. 143–68.

Birks, H. J. B. (1974) Numerical zonations of Flandrian pollen data. *New Phytol.* 73, 351–8.

Bonny, A. P. (1972) A method for determining absolute pollen frequencies in lake sediments. *New Phytol.* 71, 393–405.

Brooks, D. and Thomas, K. W. (1967) The distribution of pollen grains on microscope slides. I. The non-randomness of the distribution. *Pollen Spores.* 9, 621–9.

Brooks, J. and Shaw, G. (1968) Identity of sporopollenin with older kerogen and new evidence for the possible biological source of chemicals in sedimentary rocks. *Nature, Lond.* 220, 678–9.

Brooks, J. and Shaw, G. (1971) Recent developments in the chemistry, biochemistry, geochemistry and post-tetrad ontogeny of sporopollenins derived from pollen and spore exines. In *Pollen: Development and Physiology,* ed. Heslop-Harrison, J. Butterworth, London, pp. 99–114.

Cerceau, M. T. (1959) Clé de détermination d'Ombellifères de France et d'Afrique du Nord d'après leurs grains de pollen. *Pollen Spores* 1, 145–90.

Chamberlain, A. C. and Chadwick, R. C. (1972) Deposition of spores and other particles on vegetation and soil. *Ann. appl. Biol.* 71, 141–58.

Chambers, T. C. and Godwin, H. (1961) The fine structure of the pollen wall of *Tilia platyphyllos*. *New Phytol.* 60, 393–9.

Chambers, T. C. and Godwin, H. (1971) Scanning electron microscopy of *Tilia* pollen. *New Phytol.* 70, 687–92.

Chanda, S. (1962) On the Pollen morphology of some Scandinavian Caryophyllaceae. *Grana Palynol.* 3, 67–89.

Christensen, B. B. (1946) Measurement as a means of identifying fossil pollen. *Danm. geol. Unders.* IV (3) 2, 1–22.

Christie, A. D. and Ritchie, J. C. (1969) On the use of isenotropic trajectories in the study of pollen transports. *Naturaliste Can.* 96, 531–49 .

Clymo, R. S. (1965) Experiments on breakdown of *Sphagnum* in two bogs. *J. Ecol.* 53, 747–58.

Conolly, A. P. (1961) Some climatic and edaphic indications from the late-glacial flora. *Proc. Linn. Soc. Lond.* 172, 56–62.

Conolly, A. P. and Dahl, E. (1970) Maximum summer temperature in relation to the modern and Quaternary distributions of certain arctic-montane species in the British Isles. In *Studies in the Vegetational History of the British Isles,* ed. Walker, D. and West, R. G. Cambridge University Press, London, pp. 159–223.

Conway, V. M. (1947) Ringinglow Bog, Nr. Sheffield. *J. Ecol.* 34, 149–81.

Conway, V. M. (1954) Stratigraphy and pollen analysis of the southern Pennine blanket peats. *J. Ecol.* 42, 117–47.

Coope, G. R. (1970) Climatic interpretations of Late-Weichselian coleoptera from the British Isles. *Rev. Geogr. Phys. et Géol. Dynam.* 12, 149–55.

Coulson, C. B., Davies, R. I. and Lewis, D. A. (1960) Polyphenols in plant, humus and soil. *J. Soil Sci.* 11, 20–44.

Cox, C. B., Healey, I. N. and Moore, P. D. (1976). *Biogeography: an Ecological and Evolutionary Approach.* (2nd edition). Blackwell, Oxford.

Craig, A. J. (1972) Pollen influx to laminated sediments: a pollen diagram from northeastern Minnesota. *Ecology,* 53, 46–57.

Cushing, E. J. (1964) Redeposited pollen in Late Wisconsin pollen spectra from east-central Minnesota. *Am. J. Sci.* 262, 1075–88.

Cushing, E. J. (1967a) Evidence for differential pollen preservation in late Quaternary sediments in Minnesota. *Rev. Palaeobotan. Palynol.* 4, 87–101.

Cushing, E. J. (1967b) Late-Wisconsin pollen stratigraphy and the glacial sequence in Minnesota. In *Quaternary Palaeoecology,* ed. Wright, H. E. and Cushing, E. J. Yale University Press, New Haven and London, pp. 59–88.

Dahl, E. (1951) On the relation between summer temperature and the distribution of alpine vascular plants in the lowlands of Fennoscandia. *Oikos* 3, 22–52.

Dale, M. B. and Walker, D. (1970) Information analysis of pollen diagrams. I. *Pollen Spores* 12, 21–37.

Davies, R. I., Coulson, C. B. and Lewis, D. A. (1964) Polyphenols in plant, humus and soil. *J. Soil Sci.* 15, 299–317.

Davis, M. B. (1961) The problem of rebedded pollen in late-glacial sediments at Taunton, Massachusetts. *Am. J. Sci.* 259, 211–22.

Davis, M. B. (1963) On the theory of pollen analysis. *Am. J. Sci.* 261, 897–912.

Davis, M. B. (1965) A method for determination of absolute pollen frequency. In *Handbook of Palaeontological Techniques,* ed. Kummel, B. G. and Raup, D. M. Freeman, San Francisco.

Davis, M. B. (1966) Determination of absolute pollen frequency. *Ecology* 47, 310–11.

Davis, M. B. (1967) Pollen deposition in lakes as measured by sediment traps. *Geol. Soc. Am. Bull.* 78, 849–58.

Davis, M. B. (1968) Pollen grains in lake sediments: redeposition caused by seasonal water circulation. *Science* 162, 796–9.

Davis, M. B. (1969) Climatic changes in southern Connecticut recorded by pollen deposition at Rogers Lake. *Ecology* 50, 409–22.

Davis, M. B., Brubaker, L. B. and Beiswenger, J. M. (1971) Pollen grains in lake sediments: pollen percentages in surface sediments from Southern Michigan. *Quaternary Res.* 1, 450–67.

Davis, M. B. and Deevey, E. S. (1964) Pollen accumulation rates: estimates from late-glacial sediment of Rogers Lake. *Science* 145, 1293–5.

Deevey, E. S. and Potzger, J. E. (1951) Peat samples for radiocarbon analysis: problems in pollen statistics. *Am. J. Sci.* 249, 473–511.

Dickson, C. A. (1970) The study of plant macrofossils in British Quaternary deposits. In *Studies in the Vegetational History of the British Isles,* ed. Walker, D. and West, R. G. Cambridge University Press, London, pp. 233–54.

Dickson, J. H. (1973) *Bryophytes of the Pleistocene.* Cambridge University Press, London.

Dimbleby, G. W. (1952) The historical status of moorland in north-east Yorkshire. *New Phytol.* 51, 349–54.

Dimbleby, G. W. (1957) Pollen analysis of terrestrial soils. *New Phytol.* 56, 12–28.

Dimbleby, G. W. (1961) Soil pollen analysis. *J. Soil Sci.* 12, 1–11.

Dimbleby, G. W. (1962) The development of British heathlands and their soils. *Oxford. For. Mem.* 23.

Dimbleby, G. W. and Speight, M. C. D. (1969) Buried soils. *Adv. Sci.* 26, 203–6.

Echlin, P. (1968) Pollen. *Scientific American* 218, 80–90.

Echlin, P. and Godwin, H. (1968) The ultrastructure and ontogeny of pollen of *Helleborous foetidus* L. I.

The development of the tapetum and Ubisch bodies *J. Cell Sci.* 3, 175–86.

Engstrom, D. R. and Maher, L. J. (1972) A new technique for volumetric sampling of sediment cores for concentrations of pollen and other microfossils. *Rev. Palaeobotan. Palynol.* 14, 353–7.

Erdtman, G. (1931) The Boreal hazel forests and the theory of pollen statistics. *J. Ecol.* 19, 158–63.

Erdtman, G. (1943) Pollenspektra från svenska växtsamhallen jämte pollenanalytiska markstudier i södra Lappland. *Geol. Fören Stockh. Förhandl.* 65, 37–66.

Erdtman, G. (1956) 'LO analysis' and 'Welcker's rule'. A Centenary. *Svensk. bot. Tidskr.* 50, 135–41.

Erdtman, G. (1960) The acetolysis method. *Svensk. bot. Tidskr.* 54, 561–64.

Erdtman, G. (1963) Palynology, *Adv. in Botanical Research* 1, 149–208.

Erdtman, G. (1966) Sporoderm morphology and morphogenesis. A collocation of data and suppositions. *Grana Palynol.* 6, 318–23.

Erdtman, G. (1969) *Handbook of Palynology. Morphology. Taxonomy. Ecology.* Munksgaard, Copenhagen.

Erdtman, G., Berglund, B. and Praglowski, J. (1961) *An Introduction to a Scandinavian Pollen Flora. Vol. I.* Almqvist and Wicksell, Stockholm.

Erdtman, G. and Erdtman, H. (1933) The improvement of pollen analysis technique. *Svensk. bot. Tidskr.* 27, 347.

Erdtman, G., Praglowski, J. and Nilsson, S. (1963) *An Introduction to a Scandinavian Pollen Flora. Vol. II.* Almqvist and Wicksell, Stockholm.

Faegri, K. (1956) Recent trends in palynology. *Bot. Rev.* 22, 639–64.

Faegri, K. and Deuse, P. (1960) Size variation in pollen grains with different treatments. *Pollen Spores* 2, 293–8.

Faegri, K. and Iversen, J. (1964) *Textbook of Pollen Analysis.* 2nd edition. Blackwell, Oxford.

Faegri, K. and Iversen, J. (1974) *Textbook of Pollen Analysis.* 3rd edition by Faegri, K., Blackwell, Oxford.

Faegri, K. and Ottestad, P. (1948) Statistical problems in pollen analysis. *Univ. Bergen, Arbok, Naturvitensk., vekka 1948.* No. 3, 1–29.

Ferguson, I. K. (1972) Notes on the pollen morphology of *Saxifraga nathorstii* and its putative parents, *S. aizoides* and *S. oppositifolia* (Saxifragaceae). *Kew Bulletin* 27, 475–83.

Ferguson, I. K. and Webb, D. A. (1970) Pollen morphology in the genus *Saxifraga* and its taxonomic significance. *Bot. J. Linn. Soc.* 63, 295–312.

Firbas, F. (1934) Über die Bestimmung der Walddichte und der Vegetation waldloser Gebiete mit Hilfe der Pollenanalyse. *Planta*, 22, 109–45.

Firbas, F. (1949). *Spät- und Nacheiszeitliche Waldegeschichte Mitteleuropas nördlich der Alpen. Bd. 1. Allgemeine Waldgeschichte.* Gustav Fischer, Jena.

Flenley, J. R. (1973) The use of modern pollen rain samples in the study of the vegetational history of tropical regions. In *Quaternary Plant Ecology*, ed. Birks, H. J. B. and West, R. G. Blackwell, Oxford, pp. 131–42.

Fredskild, B. (1967) Palaeobotanical investigations at Sermermuit, Jakobshavn, West Greenland. *Meddr. Grønland.* 178 (4), 1–54.

Geiger, R. (1965). *The Climate near the Ground.* University of Harvard Press, Cambridge, Mass.

Godwin, H. (1934) Pollen analysis, an outline of problems and potentialities of the method. *New Phytol.* 33, 278–305.

Godwin, H. (1940) Pollen analysis and forest history of England and Wales. *New Phytol.* 39, 370–400.

Godwin, H. (1960) The history of weeds in Britain. In *The Biology of Weeds*, ed. Harper, J. L. Blackwell, Oxford, pp. 1–10.

Godwin, H. (1968) The origin of the exine. *New. Phytol.* 67, 667–76.

Godwin, H. (1975) *History of the British Flora.* 2nd edition. Cambridge University Press, London.

Godwin, H. and Mitchell, G. F. (1938) Stratigraphy and development of two raised bogs near Tregaron, Cards. *New Phytol.* 37, 425–54.

Godwin, H., Walker, D. and Willis, E. H. (1957) Radiocarbon dating and post-glacial vegetational history: Scaleby Moss. *Proc. R. Soc. Lond.* B. 147, 352–66.

Godwin, H. and Willis, E. H. (1959) Radiocarbon dating of the late-glacial period in Britain. *Proc. R. Soc. Lond.* B. 150, 199.

Gordon, A. D. and Birks, H. J. B. (1972) Numerical methods in Quaternary palaeoecology. I. Zonation of pollen diagrams. *New Phytol.* 71, 961–79.

Gordon, A. D. and Birks, H. J. B. (1974) Numerical methods in Quaternary palaeoecology. II. Comparison of pollen diagrams. *New Phytol.* 73, 221–49.

Gray, J. (1965) Extraction techniques. In *Handbook of Palaeontological Techniques*, ed. Kummel, B. G. and Raup, D. M. Freeman, San Francisco, pp. 530–87.

Green, B. H. (1968) Factors influencing the spatial and temporal distribution of *Sphagnum imbricatum*. Hornsch. ex Russ. in the British Isles. *J. Ecol.* 56, 47–58.

Gregory, P. H. (1961) *The Microbiology of the Atmosphere.* Leonard Hill, London.

Hafsten, U. (1961) Pleistocene development of vegetation and climate in the Southern High Plains, as evidenced by pollen analysis. In *Palaeoecology of the Llano Estacado*, ed. Wendorf, W., Museum of New Mexico Press, Publ. 1, Fort Burgwin Research Centre, pp. 59–91.

Hansen, H. P. (1949) Pollen content of moss polsters in relation to forest composition. *Am. midl. Naturalist* 42, 473–9.

Havinga, A. J. (1964) Investigation into the differential corrosion susceptibility of pollen and spores. *Pollen Spores* 4, 621–35.

Hebda, R. J. and Lott, J. N. A. (1973) Effects of different temperatures and humidities during growth on pollen morphology; an SEM study. *Pollen Spores* 15, 563–71.

Heslop-Harrison, J. (1968) Pollen wall development. *Science* 161, 230–8.

Heslop-Harrison, J. (1971a) The pollen wall: structure and development. In *Pollen: Development and Physiology*, ed. Heslop-Harrison, J. Butterworth, London, pp. 75–98.

Heslop-Harrison, J. (1971b) Sporopollenin in the biological context. In *Sporopollenin*, ed. Brooks, J. B., Grant, P. R., Muir, M., Van Gijzel, P. and Shaw, G. Academic Press, London and New York, pp. 1–30.

Heslop-Harrison, J. and Dickinson, H. G. (1968) Exine formation in the pollen grain. *Nature Lond.* 220, 926–7.

Heusser, C. J. (1969) Modern pollen spectra from the Olympic Peninsula, Washington. *Bull. Torrey bot. Club.* 96, 407–17.

Hibbert, F. A., Switsur, V. R. and West, R. G. (1971) Radiocarbon dating of Flandrian pollen zones at Red Moss, Lancashire. *Proc. R. Soc. Lond.* B. 177, 161–76.

Hicks, S. P. (1971) Pollen analytical evidence for the effect of prehistoric agriculture on the vegetation of north Derbyshire. *New Phytol.* 70, 647–68.

Hirst, J. M. (1952) An automatic volumetric spore trap. *Ann. Appl. Biol.* 39, 257–65.

Hirst, J. M., Stedman, O. J. and Hogg, W. H. (1967a) Long-distance spore transport: methods of measurement, vertical spore profiles and the detection of immigrant spores. *J. gen. Microbiol.* 48, 329–55.

Hirst, J. M., Stedman, O. J. and Hurst, G. W. (1967b) Long-distance spore transport: vertical sections of spore clouds over the sea. *J. gen. Microbiol.* 48, 357–77.

Holdgate, M. W. (1955) The vegetation of some British upland fens. *J. Ecol.* 43, 389–403.

Hyde, H. A. (1952) Studies in atmospheric pollen. A six year census of pollens caught at Cardiff 1943–8. *New Phytol.* 51, 281–93.

Hyde, H. A. (1969) Aeropalynology in Britain—an outline. *New Phytol.* 68, 579–90.

Iversen, J. (1941) Landnam i Danmarks Stenalder. *Danm. geol. Unders. IIR.* Nr. 66, 1–67.

Iversen, J. (1944) *Viscum, Hedera* and *Ilex* as climatic indicators. *Geol. För. Stockh. Förh.* 66, 463–83.

Iversen, J. (1964) Retrogressive vegetational succession in the post-glacial. *J. Ecol.* 52 (*Jubilee Symp. Suppl.*) 59–70.

Janssen, C. R. (1959) *Alnus* as a disturbing factor in pollen diagrams. *Acta Bot. Neerl.* 8, 55–8.

Janssen, C. R. (1966) Recent pollen spectra from the deciduous and coniferous–deciduous forests of northeastern Minnesota: a study in pollen dispersal. *Ecology,* 47, 804–25.

Janssen, C. R. (1967a) A postglacial pollen diagram from a small *Typha* swamp in Northwestern Minnesota, interpreted from pollen indicators and surface samples. *Ecol. Monogr.* 37, 145–72.

Janssen, C. R. (1967b) A comparison between the recent regional pollen rain and the subrecent vegetation in four major vegetation types in Minnesota (USA). *Rev. Palaeobotan. Palynol.* 2, 331–42.

Janssen, C. R. (1973) Local and regional pollen deposition. In *Quaternary Plant Ecology*, ed. Birks, H. J. B. and West, R. G. Blackwell, Oxford, pp. 31–42.

Jørgensen, S. (1967) A method of absolute pollen counting. *New Phytol.* 66, 489–93.

Jowsey, P. C. (1966) An improved peat sampler. *New Phytol.* 65, 245–8.

King, J. E. and Kapp, R. O. (1963). Modern pollen rain studies in eastern Ontario. *Can. J. Bot.* 41, 243–52.

Knox, A. S. (1942) The use of bromoform in the separation of non-calcareous microfossils. *Science,* 95, 307.

Knox, E. M. (1951) Spore morphology in British ferns. *Trans. Bot. Soc. Edinb.* 35, 437–49.

Lambert, J. M., Jennings, J. N., Smith, C. T., Green, C. and Hutchinson, J. N. (1961) *The Making of the Broads.* A reconsideration of their origin in the light of new evidence. Royal Geographical Society Research Series No. 3. London.

Leffingwell, H. A. and Hodgkin, N. (1971) Techniques for preparing fossil palynomorphs for study with the scanning and transmission electron microscopes. *Rev. Palaeobotan. Palynol.* 11, 177–99.

Lewis, D. M. and Ogden, E. C. (1965) Trapping methods for modern pollen rain studies. In *Handbook of Palaeontological Techniques*, ed. Kummel, B. G. and Raup, D. M. Freeman, San Francisco, pp. 613–26.

Lichti-Federovich, S. and Ritchie, J. C. (1965) Contemporary pollen spectra in Canada. II. The forest–grassland transition in Manitoba. *Pollen Spores* 7, 63–87.

Lichti-Federovich, S. and Ritchie, J. C. (1968) Recent pollen assemblages from the western interior of Canada. *Rev. Palaeobotan. Palynol.* 7, 297–344.

Lieux, M. H. (1972) A melissopalynological study of 54 Louisiana (USA) honeys. *Rev. Palaeobotan. Palynol.* 13, 95–124.

Mackereth, F. J. H. (1958) A portable core-sampler for lake deposits. *Limnol. Oceanogr.* 3, 181–91.

Maher, L. J. (1972) Nomograms for computing 0·95 confidence limits of pollen data. *Rev. Palaeobotan. Palynol.* 13, 85–93.

Mason, H. L. (1949) Evidence for the genetic submergence of *Pinus remorata*. In *Genetics, Palaeontology and Evolution*, ed. Jepsen, G. L., Simpson, G. G. and Mayr, E. Princeton University Press, Princeton, N.J., pp. 356–62.

Matthews, J. (1969). The assessment of a method for the determination of absolute pollen frequencies. *New Phytol.* 68, 161.

McAndrews, J. H., Berti, A. A. and Norris, G. (1973) *Key to the Quaternary Pollen and Spores of the Great Lakes Region*. Royal Ontario Museum, Toronto.

McAndrews, J. H. and Wright, H. E. (1969) Modern pollen rain across the Wyoming basins and the northern Great Plains (USA). *Rev. Palaeobotan. Palynol.* 9, 17–44.

McVean, D. N. and Ratcliffe, D. A. (1962) *Plant Communities of the Scottish Highlands*. Monographs of the Nature Conservancy No. 1. HMSO, London.

Mitchell, G. F. (1942) A composite pollen diagram from Co. Meath, Ireland. *New Phytol.* 41, 257–61.

Mitchell, G. F. (1956) Post-Boreal pollen diagrams from Irish raised bogs. *Proc. R. Irish Acad.* B. 57, 185–251.

Moar, N. T. (1969) Possible long-distance transport of pollen to New Zealand. *N.Z. J. Botany* 7, 424–6.

Moar, N. T. (1970) Recent pollen spectra from three localities in the South Island, New Zealand. *N.Z. J. Botany* 8, 210–21.

Moe, D. (1974) Identification key for trilete microspores of Fennoscandian pteridophyta. *Grana* 14, 132–42.

Moore, P. D. (1968) Human influence upon vegetational history in North Cardiganshire. *Nature, Lond.* 217, 1006–9.

Moore, P. D. (1970) Studies in the vegetational history of mid-Wales. II. The late-glacial period in Cardiganshire. *New Phytol.* 69, 363–75.

Moore, P. D. (1972) Studies in the vegetational history of mid-Wales. III. Early Flandrian pollen data from west Cardiganshire. *New Phytol.* 71, 947–59.

Moore, P. D. (1973) The influence of prehistoric cultures upon the initiation and spread of blanket bog in upland Wales. *Nature, Lond.* 241, 350–3.

Moore, P. D. and Bellamy, D. J. (1974) *Peatlands.* London.

Moore, P. D. and Chater, E. H. (1969a) The changing vegetation of west-central Wales in the light of human history. *J. Ecol.* 57, 361–79.

Moore, P. D. and Chater, E. H. (1969b) Studies in the vegetational history of mid-Wales. I. The post-glacial period in Cardiganshire. *New Phytol.* 68, 183–96.

Mosimann, J. E. (1962) On the compound multinomial distribution, the multivariate β distribution and correlations among proportions. *Biometrika* 49, 65–82.

Mosimann, J. E. (1963) On the compound negative multinomial distribution and correlations among inversely sampled pollen counts. *Biometrika* 50, 47–54

Mosimann, J. E. (1965) Statistical methods for the pollen analyst. Multinomial and negative multinomial techniques. In *Handbook of Palaeontological Techniques*, ed. Kummel, B. G. and Raup, D. M. Freeman, San Francisco, pp. 636–73.

Mothes, K., Arnoldt, C. and Redmann, H. (1937) Zur Beständesgeschichte ostpreussischer Wälder. *Schr. Phys-ökon. Ges. Königsh.* 49.

Oldfield, F. (1959) The pollen morphology of some of the West European Ericales. *Pollen Spores* 1, 19–48.

Oldfield, F. (1965) Problems of mid-post-glacial pollen zonation in part of north-west England. *J. Ecol.* 53, 247–60.

Oldfield, F. (1970) Some aspects of scale and complexity in pollen-analytically based palaeoecology. *Pollen Spores* 12, 163–71.

O'Sullivan, P. E. (1973) Contemporary pollen studies in a native Scots pine ecosystem. *Oikos* 24, 143–50.

Pearsall, W. H. (1950) *Mountains and Moorlands*. Collins, London.

Pearson, M. C. (1960) Muckle Moss, Northumberland. *J. Ecol.* 48, 647–66.

Peck, R. M. (1972) Efficiency tests on the Tauber trap used as a pollen sampler in turbulent water flow. *New Phytol.* 71, 187–98.

Peck, R. M. (1974) A comparison of four absolute pollen preparation techniques. *New Phytol.* 73, 567–87.

Pennington, W. (1974) *The History of British Vegetation*. (2nd edition) Hodder and Stoughton Educational, London.

Pennington, W. and Bonny, A. P. (1970). Absolute pollen diagrams from the British late-glacial. *Nature, Lond.* 226, 871–2.

Pennington, W., Haworth, E. Y., Bonny, A. P. and Lishman, J. P. (1972) Lake sediments in northern Scotland. *Phil. Trans. R. Soc.* B. 264, 191–294.

Phillips, L. (1972) An application of fluorescence

microscopy to the problem of derived pollen in British Pleistocene deposits. *New Phytol.* 71, 755–62.

Pigott, C. D. and Walters, S. M. (1954) On the interpretation of the discontinuous distributions shown by certain British species of open habitats. *J. Ecol.* 42, 95–116.

Proctor, M. C. F. and Lambert, C. A. (1961) Pollen spectra from recent *Helianthemum* communities. *New Phytol.* 60, 21–6.

Proctor, M. C. F. and Yeo, P. F. (1973) *The Pollination of Flowers.* Collins, London.

Ranson, J. F. and Leopold, E. B. (1962) The standard rain gauge as an efficient sampler of air-borne pollen and spores. *Pollen Spores* 4, 373.

Reitsma, T. (1966) Pollen morphology of some European Rosaceae. *Acta Bot. Neerl.* 15, 290–307.

Reitsma, T. (1969) Size modification of pollen grains under different treatments. *Rev. Palaeobotan. Palynol.* 9, 175–202.

Reitsma, T. (1970) Suggestions towards unification of descriptive terminology of angiosperm pollen grains. *Rev. Paleobot. Palynol* 10, 39–60.

Ritchie, J. C. and Lichti-Federovich, S. (1963) Contemporary Pollen spectra in central Canada. I. Atmospheric samples at Winnipeg, Manitoba. *Pollen Spores* 5, 95–114.

Ritchie, J. C. and Lichti-Federovich, S. (1967) Pollen dispersal phenomena in arctic–subarctic Canada. *Rev. Palaeobotan. Palynol.* 3, 255–66.

Rowley, J. R. and Dahl, A. O. (1956) Modifications in design and use of the Livingstone piston sampler. *Ecology* 37, 849–51.

Rowley, J. R. (1963) Non-homogeneous sporopollenin in microspores of *Poa annua. Grana Palynol.* 3, 3–20.

Rymer, L. (1973) Modern pollen rain studies in Iceland. *New Phytol.* 72, 1367–73.

Saad, S. I. (1961) Pollen morphology and sporoderm stratification in *Linum. Grana Palynol.* 3, 109–29.

Scott, H. G. and Stojanovich, C. J. (1963) Digestion of juniper pollen by Collembola. *Florida Entomologist* 46, 189–91.

Shapiro, J. (1958) The core-freezer—a new sampler for lake sediments. *Ecology* 39, 758.

Simpson, I. M. and West, R. G. (1958) On the stratigraphy and palaeobotany of a Late-Pleistocene organic deposit at Chelford, Cheshire. *New Phytol.* 57, 239–50.

Singh, G., Chopra, S. K. and Singh, A. B. (1973) Pollen rain from the vegetation of north-west India. *New Phytol.* 72, 191–206.

Skvarla, J. J. and Larson, D. A. (1966) Fine structural studies of *Zea mays* pollen. I: cell membranes and exine ontogeny. *Am. J. Bot.* 53, 1112–25.

Smit, A. (1973) A scanning electron microscopical study of the pollen morphology in the genus *Quercus. Acta Bot. Neerl.* 22, 655–65.

Smith, A. G. (1970). The influence of Mesolithic and Neolithic man on British vegetation: a discussion. In *Studies in the Vegetational History of the British Isles,* ed. Walker, D. and West R. G. Cambridge University Press, London, pp. 81–96.

Smith, A. G. and Pilcher, J. R. (1973) Radiocarbon dates and vegetational history of the British Isles. *New Phytol.* 72, 903–14.

Snaydon, R. W. and Bradshaw, A. D. (1961) Differential response to calcium within the species *Festuca ovina,* L. *New Phytol.* 60, 219–34.

Sorsa, P. (1964) Studies on the spore morphology of Fennoscandian fern species. *Ann. Bot. Fenn.* 1, 179–201.

Southworth, D. (1974) Solubility of pollen exines. *Am. J. Bot.* 61, 36–44.

Squires, R. H. and Holder, A. P. (1970) The use of computers in the presentation of pollen data. *New Phytol.* 69, 875–83.

Stewart, J. M. and Durno, S. E. (1969) Structural variations in peat. *New Phytol.* 68, 167–83.

Suess, H. E. (1970) Bristlecone pine calibration of the radiocarbon timescale. 5200 BC to the present. In *Radiocarbon Variations and Absolute Chronology, Nobel Symposium* XII, ed. Olsson, I.U. John Wiley, New York, p. 303.

Tallis, J. H. (1964) Studies on the southern Pennine peats. II. The pattern of erosion. *J. Ecol.* 52, 333–44.

Tauber, H. (1965) Differential pollen dispersion and the interpretation of pollen diagrams. *Danm. geol. Unders. IIR.* 89, 1–69.

Tauber, H. (1967a) Investigations of the mode of pollen transfer in forested areas. *Rev. Palaeobotan. Palynol.* 3, 277–87.

Tauber, H. (1967b) Differential pollen dispersion and filtration. *Proc. Congr. int. Ass. Quatern. Res.* 7, 131–41.

Taylor, J. A. and Smith, R. T. (1972) Climatic peat—a misnomer? *Proc. 4th Int. Peat Congr. Helsinki* 1, 471–84.

Terasmae, J. (1967). Recent pollen deposition in the northeastern district of Mackenzie (Northwest Territories, Canada). *Palaeogeogr. Palaeoclimatol. Palaeoecol.* 3, 17–27.

Thomas, K. W. (1964) A new design for a peat sampler. *New Phytol.* 63, 422–5.

Troels-Smith, J. (1954) Ertebøllekultur—Bondekultur. Ertebølle Culture—Farmer Culture. *Aarbøger for Nordisk Oldkyndighed og Historie* 1953, København.

Troels-Smith, J. (1955) Karakterisering af løse jordater: Characterization of unconsolidated sediments. *Danm. geol. Unders. IVR,* 3 (10).

Troels-Smith, J. (1960) Ivy, mistletoe and elm. Climatic indicators—fodder plants. *Danm. geol. Unders.* IVR, 4, 4.

Turner, J. (1962). The *Tilia* decline: an anthropogenic interpretation. *New Phytol.* 61, 328–41.

Turner, J. (1964a) The anthropogenic factor in vegetational history. I. Tregaron and Whixall Mosses. *New Phytol.* 63, 73–90.

Turner, J. (1964b) Surface sample analyses from Ayrshire, Scotland. *Pollen Spores* 6, 583–92.

Turner, J. (1970) Post-Neolithic disturbance of British vegetation. In *Studies in the Vegetational History of the British Isles*, ed. Walker, D. and West, R. G. Cambridge University Press, London, pp. 97–116.

Tyldesley, J. B. (1973) Long range transmission of tree pollen to Shetland. I. Sampling and trajectories. *New Phytol.* 72, 175–81.

von Post, L. (1916) Om skogsträdspollen i sydsvenska torfmosselagerföljder (föredragsreferat). *Geol. För. Stock. Förh.* 38, 384–94.

Walker, D. (1966) The late Quaternary history of the Cumberland lowland. *Phil. Trans. R. Soc.* B. 251, 1–210.

Walker, D., Milne, P., Guppy, J. and Williams, J. (1968) The computer assisted storage and retrieval of pollen morphological data. *Pollen Spores* 10, 251–62.

Wang, C. W., Perry, T. O. and Johnson, A. G. (1960) Pollen dispersion of slash pine (*Pinus elliottii* Engelm.) with special reference to seed orchard management. *Silvae Genet.* 9, 78–86.

West, R. G. (1964) Inter-relations between ecology and Quaternary palaeobotany. *J. Ecol.* 52 (*Jubilee Symp. Suppl.*) 47–57.

West, R. G. (1968) *Pleistocene Geology and Biology*. Longmans, London.

Westenberg, J. (1967) Testing significance of difference in a pair of relative frequencies in pollen analysis. *Rev. Palaeobotan. Palynol.* 3, 359–69.

Whitehead, D. R. and Tan, K. W. (1969) Modern vegetation and pollen rain in Bladen County, North Carolina. *Ecology* 50, 235–48.

Wodehouse, P. P. (1935) *Pollen Grains*. McGraw-Hill, New York.

Woodhead, N. and Hodgson, L. M. (1935) Preliminary study of some Snowdonian peats. *New Phytol.* 34, 263–82.

Wright, H. E. (1967) The use of surface samples in Quaternary pollen analysis. *Rev. Palaeobotan. Palynol.* 2, 321–30.

Wright, H. E., McAndrews, J. H. and van Zeist, W. (1967) Modern pollen rain in western Iran and its relation to plant geography and quaternary vegetational history. *J. Ecol.* 55, 415–43.

Wright, H. E. and Patten, H. L. (1963) The pollen sum. *Pollen Spores* 5, 445–50.

Wright, J. W. (1953) Pollen dispersion studies: some practical applications. *J. Forestry* 51, 114–18.

Yamazaki, T. and Takeoka, M. (1962) Electron microscope investigations of the fine details of the pollen grain surface in Japanese gymnosperms. *Grana Palynol.* 3, 3–12, plates I–XVII.

Zetsche, F. (1932) Kork und cuticularsubstanzen. In *Handbuch der Pflanzen-analyse*, ed. Klein, G. Springer-Verlag, Berlin, 3, 205.

Index